Evolution has Failed

E. Norbert Smith, Ph.D.

Publishing Coordinator - Tonya Holmes Shook

Printed in the United States of America

Smith, E. Norbert   1941

ISBN 10   1463619375

ISBN 13 978-1463619374

From the rising of the sun to the place where it sets, the name of the LORD is to be praised.

(Ps 113:3)

Dedication

Everyone who wants to live a godly life in Christ Jesus will be persecuted. (2 Tim 3:12)

This book is dedicated to the thousands of high school teachers, university professors and others who have been openly ridiculed, denied tenure or fired for questioning the validity of evolution. In our increasingly anti-Christian society it is no longer prudent to admit one accepts the Genesis account of creation. Although tragic, it is not unexpected as the scripture above warns. We live in a fallen world and Christians will be persecuted. It is not an option. For centuries, scientists were told to follow the evidence no matter where it leads. That has changed. Today scientists are no longer free to follow the evidence if it leads to Intelligent Design or the Creator.

University students and the general public are told science is totally objective. That used to be the case, but is no longer true. Science has become biased. There are many sacred cows in science that must not be attacked no matter how much evidence is uncovered. By far the largest member of the sacred cow herd is evolution. Although science is powerful and has given us many good things, scientists are human and can be wrong. We all see the world though a lens tinted by our worldview. Scientists have been wrong in the past and are wrong today regarding evolution.

We are warned about this in scripture. *O Timothy, keep that which is committed to thy trust, avoiding profane and vain babblings, and oppositions of science falsely so called* (1 Tim 6:20, KJV). As King David of old said, *the heavens declare the glory of God; the skies proclaim the work of his hands. Day after day they pour forth speech; night after night they display knowledge. There is no speech or language where their voice is not heard. Their voice goes out into all the earth, their words to the ends of the world.* (Ps 19:1-4a) The meaning is obvious. Those failing to see the hand of God in creation are without excuse. *The fool says in his heart, "There is no God"* (Ps 14:1a).

Foreword

Dr. Collin Patterson, at one time the Senior Paleontologist of the British Museum of Natural History in London, gave a keynote address at the American Museum of Natural History in New York City on November 5, 1981. During his remarks he made the following statement, "One of the reasons I started taking this anti-evolutionary view, or lets call it a non-evolutionary view, was last year I had a sudden realization for over twenty years I had thought I was working on evolution in some way. One morning I woke up and something had happened in the night, and it struck me that I had been working on this stuff for twenty years and there was not one thing I knew about it. That's quite a shock to learn that one can be so misled so long. Either there was something wrong with me or there was something wrong with evolutionary theory. Naturally, I know there is nothing wrong with me, so for the last few weeks I've tried putting a simple question to various people and groups of people. The question is: "Can you tell me anything you know about evolution, any one thing, any one thing that is true?" I tried that question on the geology staff at the Field Museum of Natural History and the only answer I got was silence. I tried it on the members of the Evolutionary Morphology Seminar in the University of Chicago, a very prestigious body of evolutionists, and all I got there was silence for a long time and eventually one person said, 'I do know one thing – it ought not to be taught in high school.'"

That was a quote taken from a speech thirty years ago. For Dr. Colin Patterson at that time, evolution had failed. As a theory it explained nothing; it answered none of his questions; and gave no reasonable input into his scientific investigation. For a growing number of scientists and educators nothing has changed. The concept of evolution continues to create more questions than it answers. Sadly, in the present pro-evolution climate, those working in evolution-influenced fields who question the validity of evolution must either keep their descent to themselves or risk losing their positions where they work.

Norbert Smith is an expert when it comes to dealing with evolution and its failure as a theory to explain what is actually observed in nature and its failure to allow for open discourse about those observations. As a highly qualified scientist in the fields of zoology and physiology, Dr. Smith gives many examples that bring into question the validity of evolution theory. As a scientist who has suffered through a damaged career for simply pointing out his objections to evolution theory, his passion for this subject comes through in how well he makes his points in each chapter.

This is one of those books that contain a wealth of information that can be used for personal enlightenment, illustrations for presentations and sermon preparation, as well as giving parents teaching moments for their

children. Dr. Smith has given us a valuable tool to give an answer to those who wonder why anyone could think that evolution has failed. Read on and be convinced by irrefutable arguments that it is time to move on in the discipline of science and put this disaster on the shelf along side of all the other forgotten, misguided attempts to explain reality with man's unfounded speculations.
Dr. Steve Kern

Reference
Andrew Snelling, 1990, *The Revised Quote Book*, Creation Science Foundation, Acacia Ridge D.C., Qld 4110, Australia, p 4.

Table of Contents

Introduction

Through him all things were made; without him nothing was made that has been made. (John 1:3)

The purpose of this book is twofold. One purpose is to show how evolution has failed to account for the origin, diversity and complexity of living things. Secondly, the public and especially the Christian community need to be made aware of the overwhelming scientific evidence for design that exits in nature. The evidence is all around us and many scientists today see it and have rejected evolution, yet few people are aware of this important trend. The implications are profound and far reaching both for science and for Christians. Literally thousands of former evolutionists have rejected evolution and the number is growing at an accelerating rate. In the not-to-distant future evolution will be nothing more than a footnote in the history of science. People will shake their heads in wonderment at how so many educated people could have been deceived by such an erroneous idea.

It is important for the reader to understand I write not as an outsider, but as an experienced research scientist and one who was denied tenure for rejecting evolution. I have published over 100 abstracts and technical research papers in international peer reviewed scientific journals in four diverse areas. I have lectured at major universities throughout the United States, Canada, Europe and South America. My research experience includes: the design and application of complex multichannel radio telemetry systems, behavioral and physiological thermoregulation of alligators and other reptiles and the cardiovascular response of wild animals to fear. I have put heart rate transmitters on more species of wild animals than anyone in the world and made some ground breaking discoveries. I introduced the concept of "passive fear" now known to be as important and widespread as the better known "fight or flight" response. The application of my research increased our understanding of Sudden Death Syndrome (SIDS) in infants and has saved countless lives. I have caught, studied and released over 200 alligators up to 750 pounds and my alligator research was featured on a BBC documentary, *"A smile for the crocodile."* Due largely to that documentary, I was invited as keynote speaker to a major international radio telemetry conference at Oxford University in England. I took three of my research students with me and two of them presented papers at the conference. I have also published numerous creation science papers including experimental results about how salt water animals may have survived the freshwater flood described in Genesis. I am therefore qualified to speak about science.

To make certain we are on the same page, let me begin by defining what I mean by evolution. University professors sometimes define evolution as simply "change." This is so vague it is meaningless. You must also understand that evolution is always defined by evolutionists as a fully natural process. If God or any intelligence played a part in the process, it is not evolution. Broadly there are two kinds of evolution: microevolution and macroevolution. The existence of one in no way proves the validity of the other. They are distinct and independent. The first is supported by scientific evidence, the other is not.

Microevolution is small inherited changes within the genetic makeup of living things. Certainly there is ample evidence of this. For example, by careful breeding and selection of certain genetic characteristics from wild dogs we now have over 200 distinct breeds of dogs from the tiny Chihuahua weighing less than 5 pounds to the English Mastiff weighing 350 pounds. Although impressive, these are changes within the originally created kinds described in Genesis. The phrase reproduces "after its kind" or some variation is used ten times in the first chapter of Genesis indicating genetic limits were established from the beginning. God knew the conditions on earth would change, especially after the flood, so each plant and animal was created with a variety of genetic traits in order to survive under different environmental conditions. Let me make it clear these changes are NOT macroevolution and do not account for the origin of new species. Even with the huge contrast in dog breeds and sizes, they are ALL still dogs, *Canis lupus familiaris*.

In sharp contrast, macroevolution is the ameba to man changes thought by evolutionists to account for the vast diversity and complexity of living things. Evolutionists also believe life arose spontaneously from nonliving material. While microevolution is well supported by observation and scientific evidence there is no persuasive supporting evidence for macroevolution. Evolution is accepted by faith and evolutionists do not even have a plausible theory to account for the origin of life from non-living material.

For those not trained in biology, let me be clear, evolution is NOT taught as a theory, but as an absolute fact. One of my professors in graduate school actually said, "There is more scientific evidence for evolution than there is for gravity." University students are repeatedly told those believing the creation fables in Genesis are ignorant and superstitious. Students believing the Bible are openly ridiculed as fools and compared to those believing the earth is flat. While I was a graduate student at Baylor University the comparative anatomy professor would ask each new class how many believed the Genesis account of Creation. The sons and daughters of many pastors and missionaries attended Baylor and more than half the class usually raised their hands. The professor remembered these students and ridiculed them relentlessly. Many of those students could not take the pressure and changed their major so they could drop the class. Others dropped out of the university, never to return.

Shortly after earning my Master's degree in Biology from Baylor I had an article published in the **Creation Research Society Quarterly** and my major professor found out about it. He was outraged and told me in no uncertain terms that if he had known I was a Creationist, he never would have allowed me to study at Baylor. Remember, these examples occurred at a major "Christian" university. The hostility is even more blatant at secular universities. Let me share another personal example.

Two weeks after my graduation with a doctorate in zoology from Texas Tech University I had another article published in the **Creation Research Society Quarterly**. My major professor found out about it and was so angry he formed a committee to annul my doctorate. They were unsuccessful, but this illustrates how strong the opposition is to anyone rejecting evolution. Sadly, few pastors and others in the Christian community realize how widespread this modern form of religious persecution in America is today. This must change. We must be informed and inform others of this travesty in academia.

Let me share one more example. This is a heartbreaking illustration of how far we have drifted from our Christian roots. Several years ago a fellow herpetologist and dear friend, Dr. Jeff Black taught biology at Oklahoma Baptist University in Shawnee,

Oklahoma. We had grabbed a few snakes together and he invited me to give my alligator research lecture in the Biology Department. I gave my talk and there were lots of favorable responses from faculty and students alike. People like alligators and I have many beautiful photographs and firsthand knowledge of these impressive reptiles.

Dr. Black also knew I was a Creationist and made arrangements for me to speak that evening to the Baptist Student Union (BSU) students. At a Baptist university students are required to attend a chapel service once a week. As a result, there are fewer BSU students than there are on many secular campuses. There were only 7 students in attendance and I shared some of the evidences for Creation and the Genesis flood as well as mentioning the lack of evidence for evolution. There were no questions and little interest. I thought nothing more about it and returned home.

Sadly, word got back to the Biology Department that Dr. Black had invited a Creationist to talk about Biblical creation on this Christian campus. As a result he was fired. Having been fired, he could not find another teaching job and his wife left him. He died of depression the next year. We were friends and he died because I spoke about creation at a Baptist university. I could give other examples, but I think the point has been made. The risks are high for anyone speaking out against evolution and accepting a Biblical worldview.

The same message comes across loud and clear from "educational" nature programs on TV, but with an interesting twist. Such programs often have statements such as, "natural selection is creative" or "nature is inventive" or "clever." This is clearly a modern form of idolatry because credit is given to something God created, i.e. natural law, instead of Him who created all things. It is no different from the children of Israel worshiping the golden image of a calf God created, yet few pastors are speaking out against this modern form of idolatry.

There is an important, but unstated principle in academia today, "Don't dis Darwin." Over three thousand university professors and public school teachers have been denied tenure or fired for doubting Saint Darwin. Some were dismissed for merely assigning readings by scientists critical of evolution. I know this to be true, for I was denied tenure at Northeastern Oklahoma State University for rejecting evolution dogma. I followed the textbooks and taught evolution in all of my life science classes, because of its importance in biology. Many of my students were actively involved in research projects with me and some published technical papers. I was awarded the "Outstanding Teacher" award twice, yet was denied tenure because I did not accept evolution. With my teaching and research career ended, I worked as an oilfield roughneck, taught electronics in a federal prison and finally became a truck driver until my recent retirement. It seems few today, even in Christian circles, are aware that this modern form of religious persecution is rampant in America where freedom of speech is supposed to be protected by the First Amendment of the constitution.

As we learn more about the complexity of living things, thousands of scientists, including many biologists and former evolutionists are recognizing evolution has no plausible explanation or mechanism to account for either the origin of life or the complexity known to exist inside every living cell. The Creation Research Society (http://creationresearch.org) has a membership of approximately 650 scientists, each one holding a Master's degree or above in a recognized field of science. They publish a quarterly scientific journal, the *Creation Research Society Quarterly* which remains the most prestigious of the various creation science periodicals. I have been a long time member of this organization and served on the board of directors. I have also published dozens of articles on a variety of subjects in the journal over several decades.

There is similar organization in Australia called Creation Ministrics International (www.Creation.com) that also publishes an excellent journal, *Journal of Creation.* I recently had an article published in that journal (Vol. 24(2) 2010). There are several other excellent creation science and Intelligent Design organizations. The Institute for Creation Research (www.icr.org) is an excellent and well known group. I taught a graduate course for that organization and remain in touch with the leaders. Discovery Institute is perhaps the most impressive new organization today dealing with Intelligent Design (www.discovery.org). Ken Ham's Answers in Genesis (www.answersingenesis.org) is another excellent group. Each of these organizations has diverse publications and growing memberships. Readers are encouraged to spend time at their websites for additional information on this important and timely topic. These groups also have newsletters to help people remain current as new evidences supporting creation or against evolution are discovered. Sadly, such viewpoints and discoveries are not mentioned by the media, in university classrooms or on public education TV programs. Failure to do so is misleading and the public has been deceived. Once again it is time for this to change. We must become informed and spread the word by informing others.

In a recent article Dr. Russell Humphreys, physicist at Sandia National Laboratories in New Mexico, estimates that there are actually approximately 10,000 practicing professional scientists in the USA alone who believe in a six-day creation. (See: www.ridgecrest.ca.us/~do_while/sage/v5i1of.html.) Some estimate more than 24,000 scientists have now rejected evolution for lack of scientific evidence. These scientists also understand that to present arguments or evidence against the accepted evolution viewpoint is dangerous and can cost them their careers. The number of "closet creationists" has grown explosively during the last two decades: yet most refuse to openly admit their doubts in evolution for fear of dire repercussions. This is strangely reminiscent to the children's story by Hans Christian Andersen, *The Emperor's New Clothes*. Everyone could see the king was naked, but only a child cried out, "But he isn't wearing anything at all!" So it is today with evolution. Many see its failure, but few are willing to admit it openly.

In spite of the career risks involved, over 3,000 scientists have signed Jerry Bergman's impressive and growing online *Darwin Skeptics* list and many have (www.rae.org/darwinskeptics.html) signed the Discovery Institute's *Dissent from Darwin* list No doubt there (ww.dissentfromdarwin.org) are other published lists, but the point is obvious that literally thousands of contemporary scientists today believe evolution has failed. They are looking elsewhere for answers regarding the important issue of origins.

As we continue to learn more about the extraordinary complexity of all living things, evolutionists have failed to even postulate a tenable explanation for how such things could have come about by genetic errors and random mutations. Mutations are destructive, NOT creative. It was for this reason the leaked radioactive materials in the recent tsunami damage at the nuclear power plant in Japan was considered a major health concern.

On a broader scale, evolutionists do not have a rational mechanism for the origin of life. This is why

Nobel Prize wining Francis Crick postulated in desperation that an intelligent alien from some distant galaxy must have "seeded" the earth with life. Obviously this still does not explain the origin of life. It merely moves the problem it to another galaxy. Evolution is rapidly losing ground for lack of supporting scientific evidence or even a plausible theory as to how it might have begun in the first place. As a result, tens of thousands of former evolutionists are abandoning evolution dogma like rats from a sinking ship. The responses of those who continue to accept evolution by faith are dissing all who reject it and claiming, "They are NOT real scientists." This unfounded nonsense reveals their desperation to hold on to a failed idea. In spite of what we have been told repeatedly, science is no longer objective. Prejudice reigns as never before in the ivory halls of academia. Evolution today is accepted not as science, but by faith.

If the supporting scientific evidence for evolution is lacking, why then was evolution so quickly and widely accepted? Why does it remain so firmly entrenched today? These are very important questions. Let's consider the answer in some detail. Many educated people as well as the general public would say evolution was accepted quickly because the evidence amassed by Darwin and others was overwhelming. Not so. Morality is and has always been at the core of the evolution acceptance. Many will find this shocking. Let's consider several examples from leading evolutionists past and present.

Thomas Huxley (1825-1895) was one of Darwin's strongest supporters and is best known as "Darwin's Bulldog." He was the most influential evolutionist of that time, in some ways even more so than Darwin. His publications are often cited in university classrooms today. According to Huxley, *I had motives for not wanting the world to have meaning. The philosopher who finds no meaning in the world is not concerned exclusively with a problem in pure metaphysics: he is also concerned to prove that there is no valid reason why he personally should not do as he wants to do--The philosophy of meaningless was essentially an instrument of liberation. The liberation we desired was liberation from a certain system of morality. We objected to the morality because it interfered with our sexual freedom.* It is obvious the reason evolution was embraced with such enthusiasm was not due to the overwhelming scientific evidence, but because it removed God and guilt from conscience.

British biologist Sir Julian Huxley (1887-1995) was, until his death, the world leader in modern evolution. When asked why the scientific community so quickly embraced evolution, he did not mention the evidence. Instead, he said, *"The reason we leapt at evolution was the idea of God interfered with our sexual mores."* Once again we see the motivation for acceptance of evolution was NOT the scientific evidence, but because it removes God from conscience.

The next quotation from him is even more to the point: *In the evolutionary pattern o thought there is no longer need or room for the supernatural. The earth was NOT created: it EVOLVED. So did all the animals and plants that inhabit it; including our human selves, mind and soul, as well as brain and body. So did religion.* There is no doubt, where Professor Huxley stood in his relation to God. We should think twice before we let his view of origins influence our interpretation of God's Word.

Well known Russian-born atheist Isaac Asimov was one of the world's most influential scientists and authors until his death in 1992. He and Carl Sagen strongly influenced people's view of science. Dr. Asimov said regarding origins, *In the beginning how did life begin? It seems quite certain that life developed. NOT AS A MIRACLE, but merely because molecules combined with each other along the line of least resistance. Life could not help forming under the conditions of the primitive earth any more than iron can help rusting in moist air.* Notice once again, how God is totally excluded from Creation. God must not even be seen even as a First Cause.

Next, we hear from the world-renowned American evolutionist of Harvard University, George G. Simpson. He was president of the American Association for the Advancement of Science (AAAS), one of the world's foremost scientific organizations. I was at the international meeting when he said the following: *Evolution is a fully natural process. Inherent in the physical properties of the universe by which life arose in the first place and by which all living things, past or present, have since developed. Organisms diversify into literally millions of species then the vast majority of those species perish and other millions take their places for eon until they too are replaced. If that is a foreordained plan, it is an oddly ineffective one...* What Professor Simpson is saying is that if this is the way God created living things, then God was stupid. He goes on to say, *A world in which man must rely on himself... is by no means congenial to the immature or the wishful thinkers...Life may conceivably be happier for some people in the other worlds of superstition. It is possible that some children are made happy by a belief in Santa Claus. But adults should prefer to live in a world of reality and reason.*

The pattern could not be more clear. The leaders of modern evolution were not motivated to accept evolution by the scientific evidence, but saw it as an effective way to exclude God from their world. There is neither room nor need for the supernatural. To them life, including man, is nothing more than matter in motion. Any need for God has been replaced by their view of science. Sadly, this is the viewpoint taught in our public schools and universities today. It is loudly touted by the popular media as well.

The Bible has a great deal to say about creation. There are over 1,500 Bible verses dealing with God as Creator and Sustainer of the world. Living things including man are NOT the result of accidents of random errors. The origin of life was a miracle. It was not a process. *For he spoke, and it came to be; he commanded, and it stood firm* (Ps 33:9). Nor can it be explained or understood by science. *By the word of the LORD were the heavens made, their starry host by the breath of his mouth* (Ps 33:6). God's Word says it best. *There is a way that seems right to a man, but its end is the way of death.* (Prov.14:12) And from King David, *The fool has said in his heart. "There is no God."* (Psalms 14:1) Many leading scientists today have become foolish in their unrelenting denial of God as Creator in spite of overwhelming scientific evidence to the contrary. Again according to Scripture: *Have nothing to do with godless myths and old wives' tales; rather, train yourself to be godly.* (1 Tim 4:7) *Although they claimed to be wise, they became fools.* (Rom 1:22) So it is today for those still clinging to failed evolution dogma. May the truth prevail! Evolution has failed. Let us now consider some of the reasons for its demise in detail.

Unless stated otherwise the NIV translation is used for scripture passages.

Evolution in disarray

For this reason God sends them a powerful delusion so that they will believe the lie. (II Th 2:11) That lie is evolution.

Charles Robert Darwin (18091882)

Charles Darwin has achieved unbelievable status throughout the scientific community. His ideology has influenced not only biology but also such diverse areas as behavioral science, education, social studies and even astronomy. Darwin Day 2009 celebrated the 150th anniversary of the publication of the **Origin of Species** and 200th year since Darwin's birth. Darwin fawning droned on for weeks in the media and many universities continued praising Saint Darwin throughout the year. He is even buried in Westminster Abbey along side British royalty and such notables as Charles Dickens, William Shakespeare and Isaac Newton. His theory of evolution by natural selection published in 1859 is said to be the most important unifying principle in all of biological science.

Does he deserve such honor? Let's take a closer look at Darwin for he was not always held in such high esteem. Consider the following reality check, **Laughing at Darwin** written by Edward Blick and previously published in **Sacred Cows of Science, no Objectivity Allowed.** It sets the record straight using a humorous, but fact filled and insightful approach.

"Introduction

We all love to laugh, its good medicine. We laughed at the Queen in Lewis Carroll's *Alice in Wonderland*, who said, "I sometimes believe in six impossible things before breakfast." The Darwinists are even more hilarious, they not only believe, but also teach more than six impossible fairy tales in their biology classes. The history of their pathetic attempt to pump life into the Lenin-like corpse of evolution is full of laughs. Let's peel back the skin of this rotten baloney and laugh at how this sausage was made.

Darwin was born into wealth, spent two years in medical school, dropping out after spending too much time drinking in bars. He had some divinity training but failed to make it as an Anglican minister. He was never a scientist but took a position as a naturalist on a ship and later wrote his racist books, ***The Origin of the Species and the Preservation of Favored Races*** and ***Descent of Man***. He was ignorant of genetics. He married his first cousin. All seven of his children either died young or had mental or physical disorders.

Without any facts, he conjured up his Pangenesis theory. He assumed that species changed to other species because all cells produced Gemmules. Gemmules supposedly arose by some kind of reaction to the environment. Each of these gemmules entered the sex cells of the sperm or egg (it must have been crowded in there), which later were transmitted to the offspring. Big problem! No one could find Darwin's imaginary Gemmules and Pangenesis died shortly after birth!

Reality Begins

In 1870, Adam Sedgewick, leading geologist of England, wrote Darwin: I read your book with more pain than pleasure. Parts I laughed at till my sides were sore; others I read with absolute sorrow, because I think them utterly false ...you deserted the true method of induction. Induction is reasoning from facts to theory, but Darwin reasoned from theory to facts, but neither he nor anyone else could find the facts! His writings were conjecture piled upon conjecture. Maybe and perhaps form the basis of his books! Darwin's writings were not science but philosophical musings. But something had to be done to keep the world believing Darwinism. In 1874, future Nazi-like preacher Ernest Haeckel tried by faking drawings of embryos (which he claimed repeated fish to reptile to mammal evolution).

Later that year fellow embryologist Wilhem His, Sr., exposed the hoax in detail in his book **Unsere Korperform**. His scholarly books are considered the foundation of modern embryology. For the past century other embryologists have denounced Haeckel's drawings as utter foolishness. "The theory of capitulation ... should be defunct today." Stephen J. Gould, (1980). Believe it or not, but Haeckel's most famous fakes in biology are used as proofs of evolution in biology books today. Some abortion doctors even use recapitulation: to convince would-be mothers to abort their baby. Why, that's not a human in your womb, you're in the first trimester, that is only a fish, or reptile in the second trimester! Don't school boards ever read these biology books? Haeckel's forgeries are like gonorrhea, a gift that keeps on giving!

The next attempt to resurrect Darwinism came in 1872, when the

British ship HMS Challenger dredged ocean sediments for four years looking for half-formed fossils. None were found, and since none had ever been found on land, the evolutionary fairy tale of the gradual production of billions of fossils in sedimentary strata was quietly set aside. The Challenger did provide a momentary hope. It dredged up some blob from the ocean floor and Darwinists leaped for joy. It was a live microbe, some kind of a missing link! They named it **Bathybuis Haeckel** after the old king of biological fakery, Ernest Haeckel. However in 1875 a chemist discovered it was not any form of life, but a chemical precipitate of sulphate of lime (gypsum). So, true to form, the discovery was carefully swept under the rug and hidden from the public.

In the meantime Darwin had returned to Jean-Baptiste Lamarck (1744-1829) who thought that giraffes developed long necks by stretching to reach those leaves on the top of trees. This theory died again in 1883, when German biologist Leopold Weisman cut off the tails of white mice in nineteen successive generations and the tails always reappeared. Similarly, after four thousand years of circumcision, Jewish men still had foreskins. More bad news for poor old Saint Darwin!

Darwin Resurrection

Who can rescue Darwinism? Quick, before the unwashed discover the emperor has no clothes. Finally in 1930, Austin H. Clark tried to plug the gap with a new theory, Zoogenesis. Clark was a well-respected Darwinist at the Smithsonian Institute. He had written books and six-hundred articles in five languages. However to his dismay, he could never find any evidence of macroevolution in animals or plants. In his 1930 book, *The New Evolution: Zoogenesis* he cited fact after fact proving macroevolution could not have occurred. He concluded therefore plants and animals must have sprung fully formed from dirt and water! The evolutionary world was stunned into silence. Clark was the Carl Sagan (or

Oprah Winfrey) of his day. He supposedly knew all the answers. Quickly they buried Clark's theory.

The next batter up was world famous geneticist Richard Goldschmidt, who attempted to come to the rescue of the embarrassed Darwinians by attempting to prove macroevolution was caused by mutations. For twenty-five years he was the godfather to millions of generations of gypsy moths. He zapped them with X-rays and chemicals. He found mutations produced nothing but deformities. No new species! He concluded rats were still rats and rabbits were still rabbits. In his 1940 book, *The Material Basis for Evolution*, Goldschmidt exploded the ammunition box of evolutionary theory. He literally tore the theory to pieces. No one knew how to answer him and they cannot answer him today. He was an honest atheist who faced the facts. But not wanting to acknowledge God, he proposed a new mechanism of evolution called *The Hopeful Monster Mechanism*. One day an alligator laid an egg and a turkey hatched out! You've got to remember boys and girls this is science!

Continued Deceit

For the next thirty years evolutionists were dazed and in turmoil because they had; 1, no proof that evolution had ever occurred, 2, no reasonable mechanism to explain evolution, and 3, zillions of missing links! They had bitter arguments among themselves about possible theories. The embarrassment of Goldschmidt's crude *Hopeful Monster Mechanism* caused Harvard's Stephen Gould in 1972 and a little later, Steven Stanley, of John's Hopkins University, to smarten up Goldschmidt's ugly theory by giving it a new name, Punctuated Equilibrium (Gould) and the even better high-fallutin scientific name, —Quantum Speciation (Stanley). But it was still a monster by any name.

The discovery in the 1950's of the DNA by Francis Crick and James Watson crushed the hopes of biological evolutionists. It provided clear evidence that every species is locked into its own coding pattern. Only variation within a kind (microevolution) can occur. Mathematicians showed the odds against forming DNA by chance were —quad-zillions and quad-zillions to one. Evolution by chance was impossible! But atheist Crick was not ready to believe in God. He dreamed up a new theory...are you ready for this? Some unknown space alien sprinkled sperm in our solar system and eventually creatures evolved on some planet (Krypton?). Then these evolved space creatures built a Noah's Ark rocket ship and after a long journey, zoomed down to the earth, to unloaded their zoo. Crick named his new theory Panspermia. This, boys and girls, is called science or...maybe a fairy tale! Now NASA's Life in Space Program believes this baloney and is spending billions of our tax dollars shooting up probes in our solar system looking for this sperm donor!

There you have it, the skeletons in Evolution's closet. The kooky theories of Pangenesis, Gemmules, Lamarkism, Zoogenesis, Hopeful

Monster Mechanism, Punctuated Equilibrium, Quantum Speciation and Panspermia are all just guesses. None were proven. They make good fodder for fairy tale writers. They are a barrel of laughs!

How can supposedly reasonable men believe this weird stuff and then try to pass it off as science, when it is really a cult religion? They've emptied out the stables and dumped it on the gullible public. Most Americans believe people with Ph.D.'s in science are unbiased, honest and seek the truth. But they are fallen creatures like the rest of humanity. They can have biases, be dishonest and seek only to further their own goals, honorable or dishonorable.

The Darwinists have a well-oiled propaganda machine to keep their true goals hidden from the taxpayers, two-thirds of whom believe in creation, and pay their salaries. Darwinists have web sites set up to deflect criticism of evolution and to further their legislative and judicial goals, which are to kill God and elevate humanism to His throne.

Darwinists try to hide their atheist religion from the majority of Americans, who believe in God. One of the Darwinist web sites has enlisted Jimmy Carter, our worst ever ex-president, to proselyte Christians and baptize them into The Church of Darwin (in the name of the unholy trinity, Darwin, Haeckel and Nietzsche?). These new converts are called theistic evolutionists. At the 1959 Darwinian Centennial Celebration, Julian Huxley's keynote address focused on the total repudiation of God. Huxley was asked why the world, a hundred years ago, leaped at Darwin's book *The Origin of the Species.* He answered it freed us from God's sexual mores! Evolution is a religion of no God!

Darwinists have given up public debates because they've lost hundreds of them in the 1970s and '80s. Why did they lose? As a participant in two of them I will tell you. They lost because they had no proof of macroevolution. Amazing!

No Proof! They usually tried old debate tricks of personal attacks on their opponents, i.e. you can't be a scientist because you believe the Bible, etc. But they lost because audiences were shocked. Shocked that the Darwinists had no proof! And they have none today!

In editorials and letters to the editor, the Darwinists produce no proofs. So they commonly try to bluff us Okie rubes with pompous statements like, "evolution has been proved as much as gravity and it is believed by all scientists." Get real...sure, and the moon is made of green cheese! It's all bluff, designed to shut up critics and convert us to their atheistic religion. Hitler and his propaganda chief Joseph Goebbels, would have been proud. You tell a lie long enough and loud enough and people will believe it!

Unfortunately, a lot of Americans have swallowed the lie, including about half of our college graduates. Our courts and media are full of Darwinists. Their bulldog, the ACLU, is working overtime to wipe

God from all of public life. Humanism over all is their goal!

Tragically the Darwinists have made great strides in wrecking western civilization. In the first half of the twentieth century, the religion of Darwin, Nietzsche and Haeckel became the religion of Hitler and his Nazi gang. The result was the murder of millions in their attempt to produce the Aryan super race and a victorious Germany. World War II was the most violent form of evolutionism ever seen. In the last half century, evolution hijacked America and its schools and inflicted a great defeat on American culture. Crime has skyrocketed, homosexuality and gay marriage have been mainstreamed, and our morals have submerged into a cesspool. Why? Kids brainwashed with this kooky nonsense are taught that they evolved from apes, so they act like apes. If it feels good, do it.

Not only are the Darwinians scrambling to deflect attacks from creationists, but also they are also arguing with each other over their different theories. So heated is the debate that one Darwinian says there are times when he thinks about going into a field with more intellectual honesty, *"the used car business."* (*Newsweek*, April 8, 1985, p. 80)

I suppose that nobody will deny that it is a great misfortune if an entire branch of science becomes addicted to a false theory. But this is what has happened in biology...I believe that one day the Darwinian myth will be ranked as the greatest deceit in the history of science (Soren Lovtrup, *The Refutation of a Myth*, 1987)."

Thank you Ed Blick...you certainly made the point that evolution was bogus from the start. In spite of these problems and many more, evolution soon gained scientific support in biology. The general public and especially Christians were much slower accepting evolution as the explanation for the origin of life and diversity of living things. Laws were even passed making it illegal to teach evolution in public schools. There was an avalanche of publicity about evolution with the infamous Scopes Trial of 1925. Although Scopes was found guilty, the verdict was overturned on a technicality and was never brought back to trial. Public opinion changed and evolution soon became a central part of biological science. Other laws were passed prohibiting the teaching of evolution, but public opinion morphed quickly after the launching of Sputnik. This was the first earth orbiting artificial satellite launched by the Soviet Union on October 4, 1957 and the arms race began in earnest. Suddenly science gained greater respect and with it support for evolution. Darwinian evolution became known as the single most important unifying principle in biology. It was no longer taught as theory, but as absolute established fact.

Science has given us many things and scientists are respected as few others. It seems people in our society accept whatever scientists say even in matters outside their areas of expertise. Many feel they must accept evolution because they have been told repeatedly this is what all scientists accept as true. It is always taught as absolute established fact. In an attempt to avoid alienation by the scientific community some Christians and church leaders have attempted to see evolution as the method God used to create. This theistic evolution view is taught in some Christian universities and ocassionally from the pulpit by pastors. Could this be true? Did God use evolution to create the living world? Some Christians think that evolution has only modified the scientific view of the supernatural Creation. Nothing could be farther from the truth! Such a view has neither the support of the scientific community nor that of Christians accepting the Bible as the innerant Word of the Living God. Evolution totally excludes any form of supernatural involvement in Creation. The two views are not compatible. They are oposing viewpoints. Scripture is very clear. Creation was a supernatural miracle. It was not a process. All kinds of living things were created during the six literal day week of Creation. The days of Creation in Genesis are clearly defined as literal twenty-four hour days with the repeated phrase, "evening and morning."

Let me close this chapter with another personal account. One summer I was invited to give my alligator seminar at the Oklahoma University field station on Lake Texoma. It was nearly the end of the summer and both students and professors commented that mine was the best research seminar they had heard all summer. The reason is simple…everyone likes alligators and I had personal experience lots of beautiful photos of these impressive reptiles.

After the seminar was over I went to the break area with a few students to get a cup of coffee before my long drive home. We talked a bit about my research and I casually mentioned that I was not convinced the evidence supported evolution and we talked a few minutes about Creation. I thought nothing of it, until I got a letter from one of the students. Her name was Nancy Long and several years later we studied sea turtles at the Cayman Islands. Someone told the professor, Dr. Charles Carpenter that I had doubts about evolution. He was outraged and spent the entire lecture period the next five days ranting about how dangerous I was to biology. I have always found that strange. If he had considered me ignorant because of my rejection of what most biologists feel is the cornerstone of biology, I could deal with that and would not have been surprised. Instead he considered me dangerious. To me this seems to imply that he may have had some underlying doubts about evolution dogma himself.

If someone else had a different idea as to how the heart beats, I would not feel threatened. Instead, I would invite them into the laboratory and challenge them to provide experimental proof of their novel idea. I have thought of this often over the decades. Again, it takes more faith to cling to evolution than it does to accept a supernatural creation. I know in Whom I believe. Perhaps Dr. Carpenter saw the failure of evolution, but was unwilling to admit it to himself or to the class.

Reference

Blick, Edward, *Laughing at Darwin* in *Sacred Cows in Science no Objectivity Allowed*, Smith, E. N., Editor, 2010. Available online or at bookstores.

Show me the evidence

O Timothy, keep that which is committed to thy trust, avoiding profane and vain babblings, and oppositions of science falsely so called. (1 Tim 6:20, KJV)

Evolution is presented as established fact to university students by their professors and textbooks and to the public by nature programs on educational TV. Yet, in spite of the alleged overwhelming scientific evidence in support of evolution, there are many learned evolution dissenters and the number is growing explosively. Modern education has failed to tell students of the thousands of former evolutionists that now reject it. Nor are students assigned readings critical of evolution dogma. This must change. Failure to do so is dishonest. Evolution has failed to live up to expectations on many fronts and students as well as the general public needs to know this and understand why it has occurred. We will look at a few of the many examples that are not supported by evolution, yet few university students hear of them. Let us begin, but hang on as it will be a rather bumpy ride.

Origin of Life

This is the greatest obstacle for evolutionists to deal with and is the major reason so many evolutionists are looking elsewhere for an explanation for the origin of living things. In Darwin's day living cells were thought to be little more than tiny bags of salt water. This simplistic view has changed dramatically the past few decades as we are learning more about living cells at the subcellular and molecular level. Without a doubt, the origin of life is an evolutionist's ultimate nightmare. Even the "simplest" bacterial cell is complex beyond comprehension and boggles the sharpest minds in science. Consider the following quotation from Nobel Prize winning microbiologist Francis Crick:

> "To produce this miracle of molecular construction all the cell need do is to string together the amino acids...in the correct order....Suppose the chain is about two hundred amino acids long; this is, if anything, rather less than the average length of proteins of all types. Since we have just twenty possibilities at each place, the number of possibilities is twenty multiplied by itself some two hundred times. This is...approximately 10 to the 260th power or one followed by 260 zeros! That number is beyond our everyday

comprehension…the number of fundamental particles (atoms, speaking loosely) in the entire visible universe is estimated to be 10 to the 80th power...quite paltry by comparison to 10 to the 280th power. Moreover we have only considered a polypeptide chain of rather modest length. Had we considered longer ones as well, the figure would have been even more immense."

…

"An honest man, armed with all the knowledge available to him now, could only state that in some sense the origin of life appears at the moment to be almost a miracle, so many are the condition which would have had to have been satisfied to get it going. (*Life itself*, Francis Crick, pp 51-52, 88)"

Certainly the proper amino acid sequence is important, but the way the protein is folded is many times more complex and every bit as important as the correct amino acid chain. Consider the following illustration.

In 1999 IBM announced the launch of a 100 million dollar project to better understand how protein is made by all living cells. The project was dubbed "Blue

Gene" and used the world's most powerful computer. The computer could perform one quadrillion computations per second. The most complex part of protein synthesis is not arranging the amino acids in the proper sequence, but is the folding of the protein molecule into the correct three dimensional configuration. Without this important step it will not function. Even a mid-sized protein such as hemoglobin is complex beyond comprehension. In order to fold the protein into the proper shape, it took this huge computer an entire year to complete. In contrast a living cell routinely accomplishes this complex task in one minute. This extraordinary complexity is the major reason evolution has failed. Such complexity could not come into existence by random chance. A Designer is required.

According to Dr. Robert Shapiro, Professor Emeritus at New York University: *While chemists have succeeded in making the molecules of life—or their components—in the lab out of simpler molecules...the tightly controlled processes in the chemistry lab can't be mistaken for what would have happened on the early earth. Any abiotically prepared replicator before the start of life is a fantasy. (Harvard Gazette 10/23/2008).*

There are even more difficult problems for the evolutionist. Mutations are errors in the transmission of genetic information from one generator to the next or from one cell the next. Errors are detrimental, they are NOT creative. This fact is well known as we are warned of the dangers of radiation because it increases the rate of mutations. Once again the failure of evolution is obvious to all. Not only is there no rational theory for the origin of life, but there is no explanation how the first living organism could become more and more complex over time resulting in the diversity of living things and the origin of man. Evolution has no answers and has failed miserably to account for the origin, complexity and diversity of living things. It is more a doctrine of blind faith than science. It is for these and many other reasons that thousands of former evolutionists have abandoned evolution dogma and are looking elsewhere for answers. Scripture has the answers to this dilemma. God alone is responsible for the origin and diversity of living things including man. It was NOT a natural process to be explained by science, but was a supernatural miracle. He alone should be honored and worshiped as Creator of all things. Again it is obvious to all who examine the evidence with an open mind. ***Through him all things were made; without him nothing was made that has been made.*** (John 1:3) Let's be intellectually honest and give credit where credit is due.

Upon entering college, biology students are hit with a barrage of evidences for evolution from their professors and textbooks. Such arguments are convincing to naïve students for several reasons. Few university students have studied the evidence for and against evolution. Most have not heard any opposing views at home, church or from the media. Perhaps the most important reason is they are at the university to learn and they feel their professors are experts on the topics they teach. Christian students especially are often taught not to question authority. Let's examine some of the evidences used to prop up failing evolution dogma.

The evidence used to support evolution falls into two broad categories: direct and indirect evidence. Most of the evidences used are indirect and have more than one interpretation. These will be discussed below. The actual direct evidence for evolution is sparse. For decades, most textbooks have discussed the same old two or three "proofs" of evolution. In fact there is no new evidence and these "proofs" are extremely weak. The most persistent "proof" is actually based on data known to be fraudulent. Perhaps the best known evidence is that of the peppered moths of Great Britain and the second is the oft-evoked pesticide tolerance of parasites and insects. The other common "proof" of evolution is the resistance of various pathogens to certain antibiotics. Each of these will be examined and easily debunked as "proving" evolution.

Peppered Moths

The peppered moth is indigenous to England and has long existed in two common morphological types: dark and light colored. There are also intergrades, but these are not mentioned in textbooks. The evolution of peppered moths is no longer seen as an icon of evolution. A few years ago, this issue was debated *ad nauseam*, and the Darwinists lost badly. It is now widely accepted the whole argument was false and based on fraudulent data. The moths shown in the infamous photos once seen in biology textbooks were actually glued to the tree. For this and other reasons, the peppered moth argument as evidence supporting evolution has disappeared from many new biology textbooks. Yet it remains a talking point for many university professors. This dead horse needs to be resurrected for completeness as many are still exposed to this fictional tale.

Following are typical photos of the two contrasting color phases often shown in biology and evolution textbooks. Again, this argument is weak, yet many evolutionists and biology professors still cling to it for lack of more substantial evidence that evolution has occurred.

Light and Dark Phases of the Peppered Moth

Like other moths, peppered moths are active at night and often rest on tree branches or trunks during the day. During the early 1800's, or so the story goes, peppered moths rested on light-colored lichens that grew on tree trunks. Nearly all the moths collected during this period were light-colored. Only a few dark colored moths were collected as they were seen as easy targets for hungry birds. In the mid-1800's, however, factories burned so much coal that soot settled over the countryside, killing the lichens and blackening the tree trunks. Light-colored moths on dark-colored trees were easily seen and eaten by birds. As a result, more of the black moths survived and produced offspring. Within fifty years, most moths in heavily polluted areas were black. After anti-pollution laws were passed in the mid 1900's, the soot gradually disappeared and the tree trunks again became lighter as lichens returned and the number of light-colored moths increased. This is one of the standard proofs given in older textbooks that evolution has actually occurred. Many biology professors still accept to this outdated and proven erroneous argument.

To accept and perpetuate this as proof of macroevolution is naïve...or worse; it is dishonest. Both the light and dark peppered moths are still peppered moths. No new specie has evolved! How mere color change can be touted as "positive proof" of evolution is astounding. Even if the above scenario were true, one can alternately see this as evidence that a wise Creator would foresee the need for color change with future pollution. Such limited genetic drift is often real, but can hardly explain the origin of a new species. God placed limits on genetic variability for he commanded that each newly created kind would reproduce only "after its kind" and current evidence continues to support it remains true today.

As described above, fraud regarding the alleged evolution of peppered moths was rampant in the original papers used to support evolution. What is shocking is those fraudulent photos appeared for decades in high school and university biology textbooks. Thankfully, they are no longer used in most current textbooks. Again, it is important to describe this evidence to illustrate the extreme weakness of the arguments used to support evolution and how very resistant many biologists to admitting the evidence was invalid. We will next examine the evidence of pesticide resistance for this, too, is often presented as a positive proof evolution is occurring.

Pesticide Resistance

I have lived most of my life on the family farm in western Oklahoma. I am well aware that flies and other pests develop resistance to various control methods. Flies and mosquitoes develop resistance to the chemicals used to control, repel or kill them. Farmers and ranchers must change to different kinds of fly repellent ear tags for cattle every few years. Yes, organisms were created with the ability to respond to environmental changes, but no new specie has magically arisen. The pests remain the same species of fly, worm or mosquito. There is absolutely no evidence of macroevolution in this common adaptive response. Such resistance is not evidence of how life arose in the first place or how millions of species of plants and animals could have evolved from that first living thing. Kinds still reproduce after their own kinds as commanded by their Creator during Creation week.

Antibiotic Resistance

Likewise, evidence abounds for antibiotic-resistant bacteria. Again, my professors taught this as proof positive for macroevolution. It is not. A few years ago, it was widely reported that a person had to have a lung removed because the strain of infecting tuberculosis bacterium had become unresponsive to every known antibiotic. No doubt, this problem will intensify in the coming decades as antibiotics are used indiscriminately and become more available globally. Still, this cannot be seen as a proof of macroevolution. Such changes, although profound and sometimes deadly, did not change the tuberculosis bacterium into anything but a more resistant tuberculosis bacterium.

As anyone can clearly see, such observations support only microevolution or minor changes within the originally created kinds. Such is the evidence from the development of antibiotic resistant bacteria for this, too, has failed to provide evidence for macroevolution or the changing of one major kind of organism to another kind. Yet evolution depends on this important, but unseen and unsupported principle. The very foundation of evolution is cracked and the cracks are growing ever wider as more of the details of the profound complexity of all living things are discovered. Solid support is lacking after 150 years of searching, by tens of thousands of evolutionists, which explains why learned scientists and are abandoning evolution in droves. They no longer see evolution as supported by the scientific evidence. It baffles me how our news media ignore or suppress this important trend in science. They along with the textbook authors and many university professors are being dishonest and misleading. It needs to stop and actual proven science should be taught in the science classrooms. After all science is supposed to be the search for truth. May it be so again. Lets return honesty and critical thinking to the hallowed halls of acedemia.

Selective Breeding

Charles Darwin spent many pages in *On the Origin of Species by Means of Natural Selection, or the Preservation of Favoured Races in the Struggle for Life* documenting changes in domestic farm animals brought about by centuries of selective breeding. Yet, a sheep remains a sheep and cows remain cows. No new species has been developed. Even the recently cloned animals remain true to their parent species. We understand much more today about genetics than was known in Darwin's day. In fact, it is argued that if the scientific community had known then what we now know about DNA and the ultra conservative way inherited traits are passed from one generation to the next, Darwin's theory of evolution would have been rejected...even laughed at, as has become the case today.

We now clearly understand that a finite number of genes govern any one trait such as milk or wool production. Once all those genes are selected, additional genetic improvement is not possible. A cow cannot become merely a milk-producing machine or a sheep a walking wool factory. They must also be able to move about, eat, sleep and reproduce. Here again, Darwin was obviously wrong. He thought genetic changes could continue without limit. This is supported by the then popular argument for the origin of the long necks of giraffes. Darwin and others accepted that by stretching the neck to reach ever higher, this longer neck trait could be passed on to the offspring. Of course, we know today that there is no mechanism by which this could be passed to the offspring. This popular notion was the result of the earlier work of the French biologist Jean-Baptiste Lamarck and was called "soft inheritance" or more commonly "the inheritance of acquired characteristics." Perhaps the strongest proof is seen in Jewish boys. For thousands of years Jewish boys have been circumcised, yet the amount of foreskin of Jewish boys at birth remains undiminished.

Darwin provided many examples he thought clearly demonstrated the inheritance of acquired characteristics. He called this Lamarckian hypothesis Pangenesis and explained it in some detail in the last chapter of his book, *Variation in Plants and Animals under Domestication* (Darwin, 1868). He defined Pangenesis as a hypothesis based on the principle that somatic cells would throw off microscopic particles called "gemmules" in response to the environment.

These gemmules would eventually make their way to the germ cells where they could pass the information about the parent cells to the next generation. Of course, no mechanism for such inheritance of acquired charactistics has been found and this idea was rejected long ago. The very inheritance Darwin proposed to power evolution prohibits change without limit. Indeed, the opposite trend is now slowly spreading in agriculture with the cloning of livestock there is the total cessation of additional genetic change.

There you have it. That is all the direct evidence there is supporting macroevolution. It is surprisingly weak and inconclusive. Alternative explanations abound. There is evidence supporting microevolution, but for the all-important proof for the origin of life or macroevolution from one kind of animal or plant to another, the evidence is totally lacking. It does not exist and never has. The entire superstructure of macroevolution is based on faith and hope and not on any actual scientific evidence. Since this is the best direct evidence supporting evolution, one can see why doubts are growing today. In a recent poll of qualified scientists with terminal degrees, over fifteen thousand have doubts about evolution…and that number continues to grow.

As a student, I found it instructive that when biology instructors were asked for proofs that evolution is still occurring, they quickly asserted that the process is much too slow to witness. Yet, when students asked why there is no fossil record of transitional plants or animals changing into something different, we were told that evolution occurs too rapidly to leave fossil evidence. Let's get real. Evolutionists cannot have it both ways. They must openly confess there is very little actual evidence supporting evolution and that it is accepted largely by faith. Our students must be told the truth. We should demand nothing less in TV science documentaries, our public school classrooms and especially in higher education. Evolution is more faith based than science based and again at the core its motivation for acceptance is a moral issue and freedom from guilt. Yet, it seems the leaders in modern evolution are highly successful in keeping this secret out of textbooks and far away from science classrooms. The media seems to be partners in this charade and blatant disregard for truth. Where are those aggressive investigative reporters when you need them? Our culture has been deceived into believing the groundless lie of evolution. Evolution today has truly become a dogma. The reason for the widespread acceptance of this lie is clearly given in scripture. *For this reason God sends them a powerful delusion so that they will believe the lie* (II Th 2:11, NIV). Today that delusion is evolution. How else can the continued acceptance of evolution by well trained scientists without any hard evidence be explained?

Indirect Evidence for Evolution

The above direct evidences of evolution were not

very convincing. Now let us consider the indirect evidences used in most biology textbooks to support evolution. As mentioned earlier, alternate interpretations can be made for these indirect evidences, but such counter arguments are never mentioned in textbooks or classroom lectures. Instead, they are presented as factual evidences of evolution. Again, we must demand honesty and full disclosure. There is no place for deception in public school and university classes...yet it continues unabated, and is supported by the media. It is time we demanded the truth be taught in our science classrooms. Open discussion on both sides of this important issue is needed both in our public schools and especially in our universities. Only by considering the alternatives can true science make progress. Students and scientists must again have the freedom to follow the evidence no matter where it leads. Such discussion would also sharpen a student's ability in critical thinking...a much needed atribute in our public school graduates.

Evidence From Comparative Anatomy

Let's begin this portion of the discussion with the evidence from comparative anatomy. It is widely used as a "proof "of evolution and is easy to comprehend. It is also simple to see how the same information can also be used just as effectively support the opposing view.

In many university biology programs, comparative anatomy is a difficult, but important and required course for biology majors. It is also the first course where most biology majors begin to accept evolution as established fact. Here, students can see the similarity in structure and function of organs and structures of many different animals. Students often dissect several vertebrate animals during the course including a shark, salamander, frog and a fetal pig or cat. Many similarities are obvious and students are taught repeatedly that such similarity of structure is proof that the animals are descended from a common ancestor. Sadly, few students have the knowledge, background or courage to question the evolutionary interpretation touted relentlessly by their professors and textbooks. Most accept the arguments and are won over to the evolution camp for the rest of their lives. Such a view raises lingering doubts for Christian students about what they were taught at home and in Sunday school. Biblical inerrancy is called into question. These doubts linger and grow with each new evolution based course and Bible study often wanes. For non-Christian students, such arguments often erect barriers and block all roads to the Cross and Salvation. For these students ONLY careful learned debate can remove such barriers to Salvation. We are to be prepared for such debate and are commanded in scripture to *Always be prepared to give an answer to everyone who asks you to give the reason for the hope that you have.* (1 Pet 3:15)

First, we will examine the information from comparative anatomy presented in lectures and textbooks and see how they can be used to support evolution. Most textbooks use one or more illustrations showing the bones of the human arm, leg of some common mammal such as a dog, horse or cow, the flipper of a whale, and wing of a bird or bat. Indeed, there is a striking similarity in the number and name of the bones and to some degree even in the shape of the bones.

As a side issue, human anatomy was extensively studied and the various organs and structures named long before the internal anatomy of other animals were thoroughly examined. As various animals were dissected and the organs and structures named, it was logical to use the names given to similar human parts. Therefore, there is a bit of circular reasoning involved when a professor today not only points out the similar structures in various animals, but also uses even the names of the structures as evidence of common ancestry. They seem to forget the anatomical parts were named to show the similarity with humans.

Students must learn the names of the bones as well as many of the structures on each bone. For beginning biology students, this can be a daunting task. As they study skeletal material, they see certain similarities in the way the bones are shaped and especially how they are connected together. Once students grasp the similarity of the bones in the various animals, it is then used as a powerful "proof" of common ancestry...of ameba to human macroevolution. The professor repeatedly uses this anatomical information to illustrate the fish to-amphibian-to-reptile-to-mammal-to-human evolution. They hear the same evolution scheme repeatedly in each biology class, by different professors using similar arguments. It can be overwhelming.

The evidence from comparative anatomy is easy to grasp. For many students, this is the most powerful evidence for evolution they have seen and they understand it. Until the course in comparative anatomy, much of the support for evolution was theoretical. This course makes the evidence for macroevolution more tangible and real. Students hear the same arguments from other professors in other classes and conclude they cannot all be wrong. Indeed, most biology graduates fully accept evolution as the only scientific explanation for the origin of all the plant and animal species and for the origin of man. There is no longer need or room for God as Creator. For many, this is in sharp contrast to what they have been taught at home and at church. It is also very different from the account of Creation in the Bible. Still, as science majors, many began to accept evolution and reject the truth of God's word. I saw this countless times for Christian science majors…and it always broke my heart. Few students have the intellectual background, motivation or time to understand that the same structures can also be interpreted in a way that supports biblical Creation, as we shall see in the illustration.

The argument from comparative anatomy is only indirect evidence and certainly not an actual proof that evolution occurred. Yet, it is dogmatically presented by biology professors as proof positive to inexperienced and naïve students. It can just as easily be seen to support an all-wise Creator. Perhaps here pastors and Sunday school teachers should be better prepared. The depiction

Human evolution (Drawings by Tonya Shook)

illustrates how easy it is to view the same evidence from comparative anatomy as supporting Creation instead of evolution.

Certainly, comparing the anatomy of various vertebrates can be seen as evidence of common ancestry. The evidence can, however, be seen as supporting the creation worldview with the same certainty. Animals contain a finite number of bones and other organs. In order to provide movement and mobility, some of these bones are used in appendages in the arms and legs. It is difficult to imagine how each structure could be totally different and unique for each different animal. God, in His infinite wisdom, used similar structures in a wide variety of animals to provide movement. It is just as logical to see the similarity of structures in comparative animal anatomy as evidence of a common designer as it is to see evidence of a common ancestor. Again, the forelimb of whales, cattle and even birds are all used for locomotion and can be seen as evidence of a common Creator. Both views fit the evidence and are equally plausible. Neither is a positive proof and both are dependent on one's worldview. Let's examine another indirect evidence of evolution. This one is also fun because like with the infamous peppered moth, the comic duo of fraud and deception make another visit to the science classroom.

Evidence From Comparative Embryology

Embryology is the portion of biology that deals with the embryo and its development from fertilization to birth or hatching. Eventually, every parent hears the dreaded question from their child, "Where do babies come from?" Embryology attempts to answer that ubiquitous question in depth. This is perhaps the oldest and best known argument used as evidence to support evolution. It is also the weakest. The concept is introduced with a series of drawings presented as the embryonic stages of various animals including man.

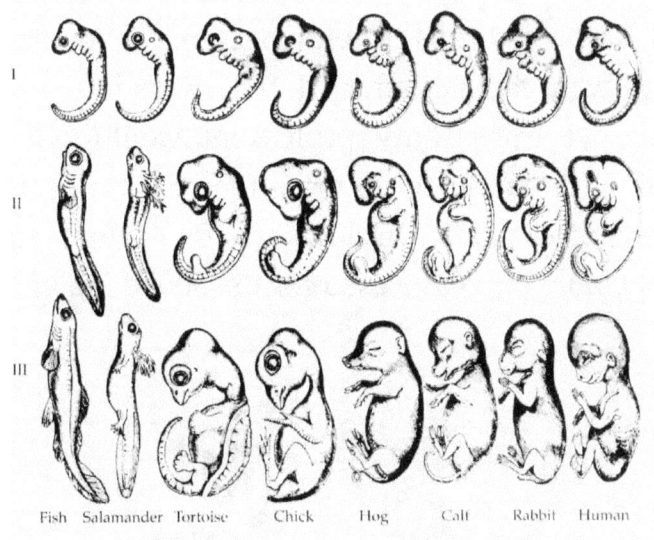

Fish Salamander Tortoise Chick Hog Calf Rabbit Human

Fraudulent Embryo Drawings credited to Ernst Haeckel

Most biology textbooks contain some form of the drawings shown above. Notice how the early embryos look alike, but become quite different as they grow and mature.

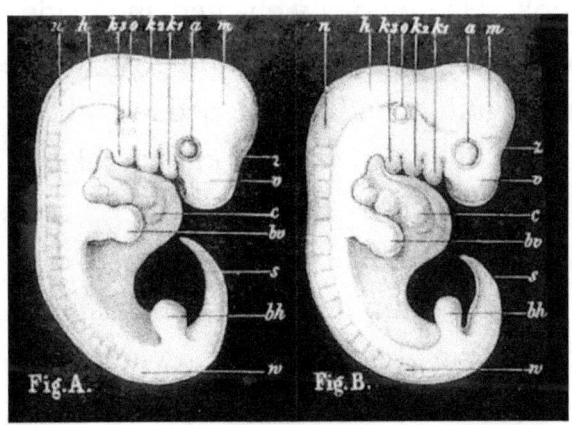

Early Human and Dog Embryo from Haeckel.

67

I must confess, the first time I saw the drawings of various early embryos I felt concern. I assumed the drawings were factual; after all, they were in my biology textbook. I trusted my professors would not use a textbook with fraudulent information. I was deceived, as were my professors. Below are the actual embryos Haeckel claimed to have used for the above drawing.

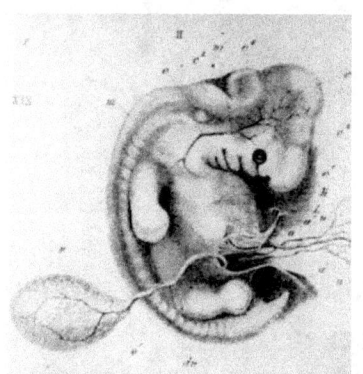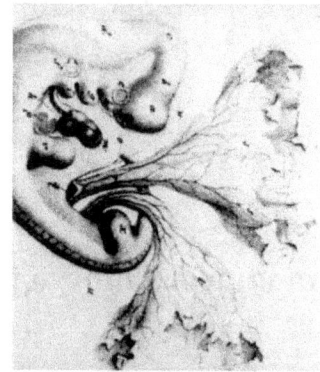

Actual human and dog embryos Haeckel said He copied.

Notice how very different the two embryos actually look. He obviously cheated and made them look more alike. He was caught cheating, tried in a court of law, found guilty and confessed to the fraudulent drawings. In spite of this, these known fake photos have been used to brainwash our students for decades. This is not right and must be stopped. Where the evidence is weak it must be admitted, but there is absolutely no excuse for using known false evidence to support evolution dogma. Again, what happened to objectivity or even honesty in science? Do scientists and university professors not have a conscience? Are such fraudlant argments the best evidence for evolution they can offer students? Apparently the answer is a resounding, 'Yes."

There is more. The evidence from comparative embryology is steeped in those beautiful words familiar to all biology majors, "ontogeny recapitulates phylogeny." It has such a beautiful sound and rolls off the tongue so effortlessly. Many biology majors find a way to drop it into dinner conversation with their parents. A typical conversation might go something like this. "Mom, please pass the potatoes and did you know ontogeny recapitulates phylogeny?" Unsuspecting parents are often so impressed with what their children are learning at the university that more money or even a new car may be forthcoming. Those useful words simply mean that during embryologic development (ontogeny) an embryo replays (or recapitulates) its own evolutionary development (phylogeny). This concept is also referred to as the Biogenic Law adding even more weight to the argument.

It is true the human heart starts beating when the baby is only two and a half weeks old. It is also true that the human heart at this stage consists of little more than a contracting tube, not unlike the heart of an adult earthworm. The human heart then progresses through stages in which it resembles the hearts of fish and amphibians. Certainly, prior to birth, the human heart is remarkably similar to the heart of most reptiles with the bulk of the blood bypassing the lungs. To unsophisticated students, the developing human heart actually does seem to be portraying its own evolutionary history in which it has developed through worm-like, fish-like and reptile-like stages. It is also true that all vertebrates begin life as a single fertilized egg and gradually become miniature adults. Thus, the early embryos actually do show certain similarities be they fish, bird or mammal. Could there be a more obvious proof of evolution? Ah, what a beautiful and useful phrase indeed.

There is much more to this simple portrayal of embryonic development than beginning students are told. There are two broad categories of organ development. Some organs, like the heart and kidneys are needed during embryonic development. They develop early and become more complex to meet the demands of the growing embryo. In contrast, other organs like the lungs are not required during embryonic development and form late in their final form. According to the Biogenic law, this would imply we evolved from animals that had no lungs. The sex organs are not needed until maturity and are not fully developed at birth. Again, the logical conclusion would be that we evolved recently from animals that lacked sex organs. Obviously, the whole argument from comparative embryology falls apart. God designed animals to develop in a logical and functional way. Organs not needed until birth or after develop late and in final form. This simple explanation can account for the similarities of certain structures during embryonic development, yet can also account for the differences. Sadly such logical opposing views are never mentioned in biology classrooms or in textbooks. Students must be told the truth including arguments contrary to the accepted viewpoint. Education is supposed to teach students to think. For that reason oposing veiws should be presented and openly discussed.

Evidence from Vestigial Organs

This has long been another of my favorite evidences for evolution, for it is easily dismissed and once again illustrates the deception common in textbooks and biology classes. This argument is also easy to comprehend. If animals actually evolved over time from one major kind of animal to another, we would expect to see some organs no longer needed as well as nascent organs or organs on the verge of becoming useful. No nascent organ has ever been described, but the argument of vestigial organs is still widely used to prop up failed evolution dogma.

Robert Wiedersheim (1848-1923) was not a good student, barely passing his final exam and his academic advancement was slow. In 1876, he became an anatomist at the University of Freiburg and soon became a comparative anatomy expert publishing several textbooks. He became widely known and respected in 1893 for publishing a list of 86 useless or vestigial organs (Wiedersheim, 1893). In his own words, he said each of those human organs had, "lost their original physiological significance." He theorized they were vestiges of past human evolution and called them "vestigial." Over the next few decades, he and others added additional organs and the list grew to 100 and then to 180 useless organs by the time of the infamous Scopes Monkey Trial. I searched diligently for the actual list of 180 organs mentioned in many textbooks, but was unsuccessful in finding such a list. This seems to be yet another myth started and widely disseminated by evolutionists.

Wiedersheim claimed the human body was a veritable walking museum of evolutionary history. He picked up and expanded on Darwin's concept of rudimentary organs as mentioned in *The Descent of Man*. Original organs included on Wiedersheim's list as useless included the human appendix, adenoids, tonsils, parathyroid gland, pineal gland, pituitary gland, thymus, valves in veins and many other important organs or tissue. As knowledge replaced ignorance, uses were found for these organs. Today, no one would claim any useless organs exist in the human body, yet this argument still appears in many textbooks and is taught by most life science professors in universities and colleges. Truth not fables must be taught in our classrooms.

Perhaps the best known of these is the human appendix. For decades, medical doctors actually accepted the evolutionary myth that the appendix was useless and during other surgical procedures, even healthy appendixes were removed, increasing surgical complications and prolonging recovery. Doctors also profited from the needless surgery.

The human appendix is a finger sized hollow tube near the end of the small intestine and entrance of the colon, and has long been taught to be a relic of our vegetarian ancestors and no longer serves a function today. Today, we know it serves many functions. Over sixty years ago we find these words from the prestigious **Quarterly Review of Biology**, "There is no longer any justification for regarding the vermiform appendix as a vestigial structure" (Straus, 1947). As a blind reservoir, it continually repopulates the colon with important bacteria lost in excrement. The walls of the appendix contain lymph tissue, known to be important in the immune response to disease. Recent evidence indicates it also lubricates oil needed for lubrication of the large intestine. Yes, I can live without my appendix or without my right arm, but this does not mean God did not give them to me for a purpose.

So it is with each of the so-called vestigial organs on the original list published by Wiedersheim. Yet, this argument is still heard in classrooms and presented in biology textbooks as yet another proof of evolution. Again, evolutionists must cling to yet another myth because the actual evidence supporting evolution is non-existent. Let us look briefly at some of those other organs originally listed as useless.

We now know the adenoids and tonsils contain lymph tissue and help in our immune response to disease.

The pituitary gland, once classified as useless, is considered the "master gland of the body" for it influences virtually every biological pathway throughout the human body. By way of the important hypothalamic-hypophysial portal system connecting our brain and body, emotions can have a profound influence on our bodily functions. The parathyroid is vital in the regulation of calcium levels and maintaining healthy bones. The thymus plays a key role in the immune response and is vital. Even the wisdom teeth have been shown necessary for proper development of the jawbone. It has long been known valves in the great veins of the body are important in returning blood to the heart. Another one that often appears on vestigial organ lists is the so-called third eyelid or nictitating membrane we all have. I don't know about yours, but mine is functional. Often, when I awake there is some sticky gunk attached to mine. My nictitating membrane functions as a sticky gunk collector and is therefore not vestigial.

It should be obvious to all there are no useless organs...only human ignorance. Our Creator knew precisely what He was doing when He made our marvelous bodies. *For this they willingly are ignorant of, that by the word of God the heavens were of old, and the earth standing out of the water and in the water.* (2 Pet 3:5, KJV) Only by willful ignorance can people deny the obvious truth of the Psalmist, *For you created my inmost being; you knit me together in my mother's womb. I praise you because I am fearfully and wonderfully made; your works are wonderful, I know that full well.* (Ps 139:13-14) Indeed, *The fool says in his heart, "There is no God." They are corrupt, their deeds are vile; there is no one who does good.* (Ps 14:1)

Evidence from Taxonomy

This is perhaps the weakest of the evidences used to prop up failed evolution dogma because it like the argument from comparative anatomy contains a huge element of circular reasoning. Taxonomy or systematics is the orderly scientific classification of plants and animals. Each plant or animal has a binomial name consisting of the genus and species. Genera are grouped into families, families into orders, orders into classes, classes into phyla and phyla into kingdoms. Perhaps an example will make it easier to comprehend. My favorite animal is the American alligator known scientifically as *Alligator mississippiensis.* Because it is Latin, it is written in italics and consists of the genus, *Alligator* and the species name *mississippiensis.* Together they make up the scientific name of the alligator common in the southeastern United States and the object of my study for decades. The only other living member of the alligator genus is the Chinese alligator, *Alligator sinensis.*

Alligators are classified in the family Alligatoridae, again with only two living alligator species. Alligators are classified in the order Crocodilia including alligators, crocodiles, caiman and a few other less known members. They belong to the class Reptilia sometimes called Sauropsida including all reptiles. This class includes not only crocodilians, but also turtles, snakes and lizards. Reptiles belong to the phylum Chordata including all animals with backbones. This also includes fish, birds and mammals. The chordates belong to the animal kingdom or Animalia. The number of kingdoms seems to be in dispute and includes the animal kingdom, plant kingdom and three other lesser known kingdoms to account for bacteria, fungi and some other organisms. The details of taxonomy are beyond the scope of this discussion and readers interested in knowing more can readily find more information in any library and many introductory biology textbooks.

Modern taxonomy is an attempt to arrange living things in a way that shows their alleged evolutionary relationships. To try to show the taxonomic relation of various animals and plants as a proof of evolution is circular reasoning because they were purposely arranged to show such a relationship. Even as a student, I saw this evidence as weak and still think it is laughable for textbooks and professors to teach it as evidence supporting evolution.

It has been argued that arranging things by similarity is a very human trait. Here is an often used example. Assume you are shipwrecked on a tropical island with ample food and water. The ship was filled with miscellaneous hardware...bolts, nuts washers and such. These items were somehow beached at the time of the shipwreck, but randomly mixed   Out of sheer boredom, many of would find ourselves going through the hardware and sorting it by size and function. Such sorting and arrangement seems a human trait. It is a way to find order in chaos.

So it is with the multitude of living organisms. Taxonomy provides such order, but is in no way is it a proof of common ancestry any more than would be a completely organized grouping of hardware implies each kind arose from a simpler bolt, nut or washer. The argument from taxonomy has absolutely no merit as an evidence of evolution.

There are a few other lesser arguments often used to support evolution. Again, there are problems with each of these. Let's end this discussion of the evidences used to support evolution with the single most important evidence, that of the fossil record. Some professors and many textbooks admit much of the preceding evidence is circumstantial at best and other interpretations are possible. Many say for the actual proof evolution, one must turn to the fossil record, the actual history of life on planet earth. Does the fossil record support evolution? You may be surprised.

Evolution and the Fossil Record

This is a perhaps the most important topic for evolutionists. Many biology textbooks adamantly teach the fossil record proves evolution has occurred. If there is indeed fossil evidence supporting evolution, then it should be taught in high school and university biology classrooms. If instead, there is little actual support for macroevolution from fossils, it must be clearly stated.

The actual fossil record is often ambiguous, supporting alternate views, yet, even the discussion of those problems and contradictions are banned from university classrooms. Let's consider the fossil evidence in some detail.

Let me begin with an actual event that will set the tone for this discussion. For decades, Duane Gish, Henry Morris, Sr. and other Creationists openly debated evolutionists about the evidence supporting evolution. Many years ago, Dr. Gish debated a leading evolutionist in England. In preparation for the debate, the evolutionist he was debating went to the British Museum of Natural History for the latest evidence. This is a world class museum with one of the best collections of fossils. The evolutionist met with the fossil curator and asked for the best fossil evidence he could use to blow Dr. Gish out of the water. The curator of fossils at the British Museum of Natural History sternly warned the evolutionist to stay far away from fossils for in truth they fail miserably to support evolution. I have long found this advice from someone who knows, most revealing, for it flies in the face of what our textbooks and professors teach students. It remains as true today as it was then. In spite of what we are told the fossil record does not support evolution.

The modern evolutionary synthesis was put together in the 1930s and 1940s by Theodosius Dobzhansky, Ernst Mayr, J.B.S. Haldane, Sewall Wright, George Gaylord Simpson, G. Ledyard Stebbins and others. But as this philosophy was fine-tuned and disseminated to the masses, it was never actually observed from the fossil evidence. I was shocked and angered when I discovered this important fact. Generations of students have been deceived by the education system. *Phyletic gradualism [gradual evolution] was an a priori assertion from the start – it was never "seen" in the rocks* (Gould & Eldredge, 1977) yet is presented as absolute fact and proof of macroevolution in countless high school and university biology textbooks and by the teachers in life science classes. When informed students challenge the aforementioned weak evidences supporting evolution, the professors reflexedly respond that for the real proof of evolution, one must only look at the fossil record. The same argument is often used in textbooks. This is central to the acceptance of evolution. Does the fossil record clearly support evolution as it is adamantly reported to do? As we shall clearly see, the fossil record fails to support macroevolution. Instead, the rich fossil record is proof positive of a global flood and rapid burial. Let's consider the fossil evidence in some detail.

The history of most fossil species includes two features particularly inconsistent with gradualism (neo-Darwinism) and their sudden appearance. *Most species exhibit no directional change during their tenure on earth. They appear in the fossil record looking pretty much the same as when they disappear; morphological change is usually limited and directionless.*

Sudden appearance.

In any local area, a species does not arise gradually by the steady transformation of its ancestors; it appears all at once and "fully formed." (Gould, 1980)

Gould was correct. Consider the following: **Both the origin of life and the origin of the major groups of animals remain unknown** (Fisher, 2003). **Almost all of our information about macroevolution has come from the fossil record and, as we have seen, there are doubts as to how reliable this is** (Palmer, 1999). **Perhaps no aspect of evolution has received such intense study as human evolution, yet this is a subject concerning which there is much debate, and about which there is still much to be learned** (Colbert, 2001). Such statements by learned secular scientists has helped the cause of ID/creation science, for non-evolutionists have been saying this since Darwin.

Skeptics of Darwin's theory have used a truly remarkable book by evolutionist Barbara J. Stahl of Saint Anselm College in New Hampshire, revealingly titled, ***Vertebrate History: Problems in Evolution.*** Sadly, this important work is out of print. Dr. Stahl, anatomy professor and paleoichthyologist, is clearly no friend of the creationist. She was, however, intellectually honest enough to write this 604-page book documenting the many problems associated with alleged evolution of the vertebrates. Darwinists were understandably quick to downplay Dr. Stahl's research. In recent years, their only "valid" criticism is that the book is dated and anything found in its pages are now passé. I strongly disagree. In 2001, Edwin H. Colbert and his coauthors published their fifth edition of *Colbert's Evolution of the Vertebrates.* Dr. Stahl's detailed research has held up all these years when compared with Colbert's more recent text. Consider carefully the following examples from her book.

Origin of Fish: "The higher fishes, when they appear in the Devonian period, have already acquired the characteristics that identify them as belonging to one or another of the major assemblages of bony or cartilaginous forms" (Stahl, 126). "Both these groups appeared in the late Silurian period, and it is possible that they may have originated at some earlier time, although there is no fossil evidence to prove this" (Colbert, 53). Contrast this lack of fossil evidence for evolution with the clear evidence for creation: the sudden appearance of fully formed vertebrates (and invertebrates!) in the fossil record. Is it any wonder that another honest evolutionist stated, *The origin of animals is almost as much a mystery as the origin of life itself* (Donoghue, 2007).

Origin of Amphibians: Since the fossil material provides no evidence of other aspects of the transformation from fish to tetrapod, paleontologists have had to speculate how legs and aerial breathing evolved (Stahl, 195). This is certainly a logical explanation of the first stages in the change from an aquatic to a terrestrial mode of life. We can only speculate about this. (Colbert, 84-85).

Origin of Snakes: The origin of the snakes is still an unsolved problem" (Stahl, 318). Unfortunately, the fossil history of the snakes is very fragmentary, so that it is necessary to infer much of their evolution. (Colbert, 154).

Origin of Birds: In the absence of fossil evidence, paleontologists can say little about the date at which these sixty-nine living families of Passeriformes appeared. (Stahl, 386) Of all the classes of vertebrates, the birds are least known from their fossil record. (Colbert, 236).

Origin of Whales: As with most tetrapods secondarily modified for aquatic living, ascertaining the terrestrial stock from which the whales came is exceedingly difficult. (Stahl, 486). Like the bats, the whales appear suddenly in early Tertiary times, fully adapted for life in the sea. (Colbert, 392). Many years ago, I had the following whimsical drawings made to help emphasize this point in my Creation lectures. Students loved them and they made a very important point...fossils do NOT support evolution.

A Whale of a Tale

There are no fossils linking whales to any other group of vertebrates.   Undaunted, evolutionists are confident whales evolved from some unknown ungulate. They would have us believe some cow-like ungulate ancestor to the whales made her way to the ocean for a refreshing swim.

She enjoyed the water and ventured farther and farther out into the sea.   Her legs were mysteriously becoming transformed into fins gliding her gleefully along.

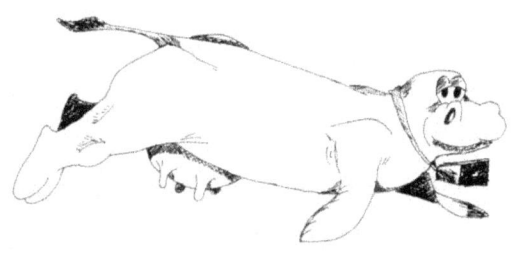

As if by magic, her legs were being modified into flippers and she learned how to hold her breath longer and longer. All the time she was very careful not to leave any fossil record of her slow transformation.

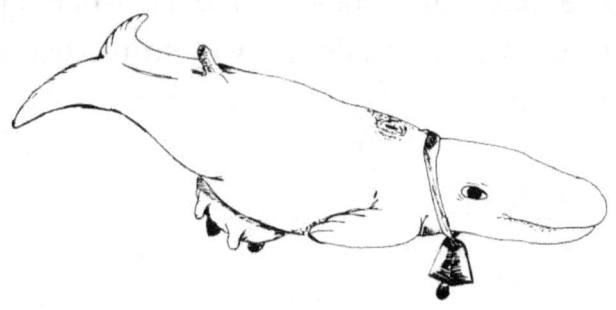

Slowly, ever so slowly, over millions of years as if by some intelligent design, she was magically transformed into a modern whale. Again, making certain no fossil evidence was left in the ubiquitous sedimentary deposits.

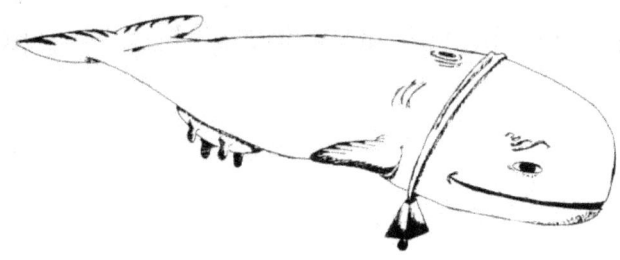

Lacking any supporting fossil evidence, I submit this theory is an udder failure, yet this fictional tale is repeated countless times in science classrooms in an attempt to prop up a failing theory. The actual fossil record fails miserably to support gradual change over time, yet that is the very essence of evolution. Evolution as become bankrupt for lack of supporting scientific evidence. Once again, I say vehemently it is the evolutionist and not the Creationist that has the greater portion of blind faith for I *know* in whom I believe.

Evolution Acceptance

The reason Darwin is so revered by the secular world today is he is thought by many to be the first to provide a plausible explanation for the origin of life, diversity of living organisms including origin of man without the need for a Creator-God. As is shown above, this premise is clearly false for there is overwhelming evidence of Creation. Indeed, as King David of old said: *The fool says in his heart, "There is no God"* (Ps 14:1a, NIV).

Let's conclude this section with what the Bible teaches. God's Word says it best. *There is a way that seems right to a man, but its end is the way of death.* (Prov.14:12) And, *The fool has said in his heart. "There is no God."* (Psalms 14:1) Many leading scientists today have indeed become foolish in their unrelenting denial of God as Creator in spite of overwhelming evidence to the contrary. The actual evidence for evolution is weak to non-existent. As a theory to explain all, it has failed miserably. Next let's examine some examples that defy an evolutionary explanation. These are truly an evolutionist's worst nightmare.

References

Ashton, Editor, 2000, *In Six Days, Why Fifty Scientists choose to believe in Creation* published by Master Books

Javor, George, 2005, *Evidences for Creation, Natural Mysteries Evolution Cannot Explain* by published by Review and Herald Publishing Company.

Averick, Moshe Rabbi, 2010, *Nonsense of a High Order, the Confused and Illusory World of the Atheist* published by Tradition and Reason Press, Chicago, Illinois.

Migrating duckweed

The name "duckweed" is misleading because it is neither a waterfowl nor an unwanted weed growing in a flowerbed. Duckweed are small floating plants found throughout much of the world where still or slow moving water abounds. Many of us have enjoyed it in our aquariums. Duckweed seldom blooms, but when it does it has the smallest flower in the world measuring only 0.3 mm or less in length. The tiny seed it produces contains an air filled cavity to facilitate flotation. Duckweed is also a high protein food and important for waterfowl. Humans in some regions of Southeast Asia eat also eat them. Duckweed contains more protein than soybeans and is rich in essential amino acids. Surprisingly, it is recently being considered a likely candidate for bio fuel and contains 5-6 times as much starch as corn. Not only does it not contribute to global warming, but it removes carbon dioxide from the atmosphere.

Duckweed is unusual for another reason. It is the only plant that migrates each year. When one thinks of the term "migration," animals such as birds, fish and perhaps a few mammals come to mind. Many of our song birds and waterfowl fly south each winter and return in the spring. Elk migrate above the timber line in summer and move down the mountains during winter. In centuries past, the American buffalo migrated during the warmer months from northern Mexico to southern Canada in herds so large it sometimes took several days for them to go past any given location. Salmon and sea turtles migrate to where they hatched to lay their eggs. Even monarch butterflies migrate each year. Still, the migration of duckweed is fascinating and presents a real nightmare for evolutionists.

Duckweed has the highest growth rate of any plant because nothing is spent or "wasted" on structural support tissue. Under ideal conditions the surface area covered by duckweed can double in less than 2 days. The Indian species, *Wolffia microscopica*, can bud off a new daughter every 30-36 hours. Thus one tiny plant could theoretically produce offspring equal to the volume of the earth in just a few months!

In temperate climates, duckweed survives the cold winter months by producing buds that sink to the bottom of the pond or migrate. They remain on the bottom in the dark cold water below the ice. In spring as the ice melts and the available light increases, oxygen is produced and buoyancy restored. Duckweed then migrates back to the surface of the pond for another year.

These tiny plants are important to the pond by not only providing food, but by shading the water resulting in lower water temperature which allows for increased dissolved oxygen. They also reduce pond water loss through evaporation by covering the surface. The migrating duckweed is unique and evolution has no rational mechanism why or how such yearly migration evolved in small steps. Perhaps even more unusual is this is the only plant that migrates. Since it is so successful for the duckweed, why did such migration not evolve in other plants? Like other complex features of living things, it is impossible to imagine how their migration could have developed in small steps by mutational error. All the parts must be in place before their role in migration can occur. Failure of the individual components to have a function would result in their elimination by natural selection. I have long seen the migrating duckweed as an example of a Creator with a sense of humor. Evolution has failed to account for the yearly migration of this tiny, but important flowering plant.

References:

Culley, D.D., Jr. et al. 1981. "Production, Chemical Quality and Use of Duckweeds (Lemnaceae) in Aquaculture, Waste Management, and Animal Feeds." *J. World Maricult. Soc.* 12 (2): 27-49.

Hillman, W.S. and D.D. Culley, Jr. 1978. "The Uses of Duckweed." *American Scientist* 66: 442-451.

Rusoff, L.L., E.W. Blakeney and D.D. Culley, Jr. 1980. "Duckweeds (Lemnaceae): A Potential Source of Protein and Amino Acids." *J. Agricult. Food Chem.* 28: 848-850.

Bombardier beetle's chemical weapon

Bombardier Beetle (Wikipedia.org)

This remarkable beetle has long been seen by many as powerful evidence for creation and continues to be a thorn in the side of evolutionists. The more we learn about this extraordinary insect and its uniquely complex defense strategy, the stronger is the evidence it was created. This is obvious to all because many unque structures and several uncommon toxic chemical agents are required before it would be of value to the beetle. Again, evolution has failed miserably to explain the origin of this powerful and effective chemical warfare armament and its accurate delivery system. It involves far too many parts and complex toxic chemicals to have come about by random mutation. Such a complicated system shouts design at many levels and is indeed testimony to an all wise Creator-God. Let's consider a few of the remarkable facts.

There are over five hundred species of bombardier beetles known for their unusual and powerful chemical defense. They can accurately fire a boiling hot foul-smelling liquid at a potential enemy. The expulsion is accompanied by a loud popping sound and provides additional protection. Bombardier beetles produce and store two powerful toxic chemicals, hydroquinone and hydrogen peroxide which collect in a reservior. These toxic chemicals are not found in any other living creature. When threatened, the two chemicals are forcefully squirted into a chamber where they are mixed with a catalytic enzyme. The resulting mixture boils and is directed to the attacking predator by a complex muscle controled valve.

Secretary cells produce hydroquinone and hydrogen peroxide, which collect in a reservoir. The reservoir opens through a muscle-controlled valve onto a thick-walled reaction chamber. This chamber is lined with cells that secrete catalyses and peroxidases. These reactions release free oxygen and generate enough heat to bring the mixture to the boiling point and vaporize about a fifth of it. Under pressure of the released gases, the valve is forced closed, and the chemicals are expelled explosively through openings at the tip of the abdomen. Each time it does this, it shoots about 70 times very rapidly. The damage caused can be fatal to attacking insects and small creatures and is painful to human skin. The flow of reactants into the reaction chamber and subsequent ejection to the atmosphere occurs cyclically at a rate of about 500 times per second and with the total pulsation period lasting for only a fraction of a second. The gland openings of some African bombardier beetles can swivel through 270° and thrust between the insect's legs, so it can be discharged in all directions with considerable accuracy.

The reason such a complex and uncommon defensive mechanism is a problem for evolutionists is obvious. All of the components, including glands to produce, store and quickly release the toxic chemicals along with the sophisticated and accurate method of mixing, aiming and expulsion must all occur together to be of any value. The development of any of the component parts without the complete system would be of no value and would be eliminated by natural selection. Such a beautifully complex weapon shouts of the wisdom of its Creator while making a laughing stock of evolution dogma. Random genetic errors could not invent such an incredibly complicated design, no matter how much time was available. Once again the Psalmist was right on. *The fool says in his heart, "There is no God." They are*

corrupt, their deeds are vile; there is no one who does good. (Ps 14:1)

References

Armitage, Mark H.; Mullisen, Luke (April 2003). *Preliminary observations of the pygidial gland of the Bombardier Beetle, Brachinus sp..* Answers in Genesis. http://www.answersingenesis.org/tj/v17/i1/beetle.asp. Retrieved 9 July 2007.

Eisner T, Aneshansley DJ (August 1999). "Spray aiming in the bombardier beetle: Photographic Evidence". *Proc. Natl. Acad. Sci. U.S.A.* **96** (17): 9705–9. doi:10.1073/pnas.96.17.9705. PMC 22274. PMID 10449758. http://www.pnas.org/content/96/17/9705.full.

Eisner T, Aneshansley DJ, Eisner M, Attygalle AB, Alsop DW, Meinwald J (April 2000). "Spray mechanism of the most primitive bombardier beetle (*Metrius contractus*)" (PDF). *J. Exp. Biol.* **203** (8): 1265–75.

Isaak, Mark (May 30 2003). "Bombardier Beetles and the Argument of Design". *TalkOrigins Archive.* http://www.talkorigins.org/faqs/bombardier.html.

Kofahl, R.E., The Bombadier Beetle shoots back, *Creation/Evolution* **2**(3):12–14, 1981.

Rue, H.M., *Bomby the Bombardier Beetle* (third printing), Institute for Creation Research, El Cajon, 1993.

"Bombardier Beetle". *Animal Facts & Photos*. Dallas Zoological Society. 2004.

http://www.dallaszooed.com/animalfacts/animalfacts.php?id=100&region=6&ci=1&li=14.

Diving Alligators

Alligators and other reptiles have lungs and breath air like we do. Alligators and other crocodilians have a unique feature that enables them to dive under water and stay submerged for hours. Having spent much of my life studying alligators, I find them uniquly interesting for many reasons. Like the aforementioned bombardier beetle crocodilians present yet another nightmare for evolutionists. Let me share some things about them that few people know. They have a simple, but unique tool that enables them to hold their breath for a long time that is interesting. It is yet another thorn in the side of evolutionists for which they have no rational argument.

All crocodilians have a completely divided ventricle meaning they have a four chambered heart like birds and mammals. They are unique and have a special feature that helps them hold their breath and remain submerged for hours. It is called the foramen of Panizza and allows blood to bypass the lungs while submerged for prolonged times. It functions much like the foramen ovale in unborn mammals including humans by reducing the work load of the heart for unborn mammals. Prior to birth mammals get their oxygen from the placenta by way of the unbilical cord. At this time it would do no good to cause all the blood to go through the lungs as it does after birth.

The foramen ovale provides a path for blood to flow from the right atrium to left atrium of the heart, bypassing the lungs. With a newborn mammal's first breath the pulmonary blood pressure decreases and the left atrial pressure exceeds the right atrial pressure closing foramen ovale forcing all the blood to flow to the lungs.

The foramen of Panizza found in crocodilians remains functional throughout their life and helps them occupy their unique ecological niche by allowing them to remain submerged where they are safe for hours at a time. Many crocodilians also have a special sphincter that prevents blood from flowing in this pathway while they are not diving. What is particularly troublesome for evolutionists is its uniqueness to crocodilians. Other diving animals such as turtles, birds, seals or even whales lack this special feature. If this useful feature evolved in crocodilians, why has it not evoloved for other diving animals? Evolutionists are once again at a loss to explain its origin in crocodilians and the lack of such a useful device in other diving animals. With the avalanche of such new information, it is becoming increasingly difficult to remain a committed evolutionist! Those pesky facts continue to stand in the way of good evolution dogma. The cracks in the foundation of evolution are widening each year as we discover more about the complexity, beauty and sheer wonder of God's creation. It requires a huge amount of blind faith to remain a devoted evolutionist. As Christians, we know in Whom we believe and it was HE that created the many marvels in all living things. ***Great is the LORD and***

most worthy of praise; his greatness no one can fathom.
(Ps 145:3)

References

Axelsson, Michael. 2001. *The crocodilian heart; more controlled than we thought?* Experimental Physiology 86:6 785-789

Axellsson, M. and C.E.Franklin 2001. The caliber of the foramen of Panizza in *Crocodylus porosus* is variable and under adrenergic control. J Comp Physiol B. 2001 May 171(4):341-6.

Davies, D., J. L. Thomas and E. N. Smith. 1982. Effect of body temperature on ventilatory control in the alligator. *J. Appl. Physiol.* 52:114-118.

Smith, E. N. 1975. Thermoregulation of the American alligator. *Physiol. Zool.* 8:117-194.

Smith, E. N. 1975. Oxygen consumption, ventilation and oxygen pulse of the American alligator during heating and cooling. *Physiol. Zool.* 48:326-337.

Smith, E. N. Editor. 2011. *Sacred Cows in Science: no Objectivity Allowed*.

Buffalo were designed to die

American Bison, another evolutionist's nightmare
Photo by the author

It is difficult to imagine an indigenous people more closely connected to an animal than were the Native Americans to the American bison or buffalo. Even today, many tribal leaders mark time from before and after the disappearance buffalo. It is estimated over 60 million bison once roamed the vast western prairies from Mexico to Canada. Their slaughter by white hunters for pelts and later only for their tongues brought near extinction to bison and marked the end of a way of life that had been in harmony with nature for thousands of years. The native people depended on the virtually unlimited bison for food, clothes, tools, medicine, ornaments and shelter. Have you ever wondered at the success of Native Americans in killing this huge animal? For decades, I marveled at how the native people could have been so successful in harvesting this important and magnificent creature. After considerable research, I now understand and the reason points clearly to the Creator who made them easy to kill in order to feed a people and it is impossible for evolutionists to explain.

Historically, there were four ways bison were killed by Native American people. Some were stampeded over cliffs or trapped in box canyons. Both of these methods were effective, but could only be used in a limited number of places and then only if the bison herd was at the right place. Neither method could reliably support the plains people for centuries.

In the sixteenth century, Spaniards and others re-introduced horses to America. It is thought over-hunting caused their earlier extinction toward the end of the ice age. Their influence on native people was immediate and profound. A third method of killing bison was running down a single animal by horse. Bison are powerful animals and it often took as many as five or more fresh horses to finally fatigue the bison.

Horses were more commonly used to kill bison as often seen in movies. A horseback rider would run alongside a running bison and shoot it with a bow and arrow. This method troubled me for years. I am not a bow hunter, but the idea of hitting the heart of a running bison from a galloping horse seemed difficult if not impossible. The answer was most unexpected and came in a graduate course in vertebrate natural history at Baylor University taught by Dr. Bryce Brown.

With the notable exception of the American bison, *all* mammals have two separate lung or pleural cavities. As we know, one side of our chest can be penetrated collapsing that lung, but the other side remains intact and that lung can support life. The bison is unique in having an incompletely divided mediastinum. There is only one functional pleural cavity containing both lungs. Thus the problem for the native bow hunter is solved. An arrow must only penetrate the chest and both lungs collapse. The fatally wounded animal would continue a few yards and die providing unlimited food, clothing and tools. Before the availability of horses, bison could be shot by stealth from a blind or other hiding place. One problem is solved, yet another serious one remains. This problem is *never* mentioned in biology or evolution classes, yet this important problem demands an answer.

The problem for the evolutionist is simple, yet revealing and difficult. Other than providing food for hungry people, of what possible selective advantage is an incompletely divided mediastinum? From an evolutionary viewpoint, it makes absolutely no sense and does *not* occur in any other mammal. Indeed, conventional wisdom would argue for the elimination of such a detrimental trait from the gene pool. Yet it remained, providing food, clothes and tools for a continent of Native Americans for millennia. It must indeed require faith and dedication to remain an evolutionist for the facts of science keep getting in the way. I know the Creator of the bison and understand this profound demonstration of love and provision for Native American people.

References

Grathwohl, K. W. and S. Derdak. 2003. *Buffalo Chest*. N. Engl. J. Med 349:19 1829.

Isenberg, A. C. 2001. *The destruction of the Bison; An Environmental History 1750-1920.* Cambridge University Press.

Krech, S. 2000. *The Ecological Indian; Myth and History*. W.W. Norton & Co. publisher.

Lewis, M and W. Clark with F. Bergon Editor. 2002. *The Journals of Lewis and Clark (Lewis and Clark Expedition)*. Penguin Classics.

Rinella, S. 2008. *American Buffalo in Search of a Lost Icon*. Spiegel; & Grau New York.

Smith, E. N. and J. Swallow. 2011 (In Press). *American Buffalo and Native Americans*. Perfect Circle Publishing Company.

Zeman, S. C. 2002. *Chronology of the American West from 23,000 B.C.E. through the Twentieth Century*, 381 pages. Page 155.

Skunk cabbage melts snow

Western skunk cabbage, *Lysichitum americanum*
Photo by the author near Snoqualmie Pass, Washington
State

Although it is not widely known even in the scientific community, several plants can produce significant amounts of metabolic heat. These include the common *Philodendron* (Seymour et. al., 1983) the sacred

107

lotus, *Nelumbo nucifera* (Seymour and Schultze-Motel, 1996) as well as the widely distributed skunk cabbage. The Western skunk cabbage, *Lysichitum americanum* is an unmistakable harbinger of spring, often making its appearance before winter's snow has gone. The common name comes from the observation that the crushed leaves smell mildly like the defensive odor of a striped skunk, *Mephitis mephitis,* but its odor is less offensive than the more widely distributed Eastern skunk cabbage. The western species is also sometimes known as the "swamp lantern," because of the way its bright yellow fluorescent flower stands out on the drab forest floor in swampy areas. *Lysichitum* is a genus of only two species, the other being *L. camtschatcensis*, which has a white spathe and is native to Northeast Asia. The plant referred to as Eastern skunk cabbage (*Symplocarpus foetidus*), has a somewhat similar appearance but a purplish or mottled brownish spathe. The Western skunk cabbage is found from Alaska south to Northern California, and east (but less common) to Montana and Idaho. It is most common west of the Cascades, but is plentiful in wet areas inland. This plant grows in swampy areas and is generally found in the wet ground under or near cedar trees. It is sometimes a dominant under story plant of cedar/alder communities, especially in areas that support Western red cedar (*Thuja plicata*).

In the past, Native Americans roasted and dried skunk cabbage, and made flour from its starch. It is believed that mainly the rootstalks and young leaves were used. However, several references report that it was a "famine food," and not a staple diet. The plant contain crystals of calcium oxalate, which sting and burn the lips and mouth if eaten raw. To be edible, the toxic leaves must be dried, pulverized, and cooked, breaking down the crystals (Pojar and MacKinnon, 1994).

In order to appreciate the remarkable physiological feature of skunk cabbage one needs to know how some vertebrates regulate their body temperature. All biochemical reactions are profoundly influenced by temperature. Rates of biochemical reactions increase with higher temperatures and slow as the temperature drops. Although there is some overlap, animals have two contrasting methods of regulating their body temperature. Warm blooded or endothermic vertebrates including mammals and birds maintain a high and constant body temperature by a combination of behavioral and physiological methods. Behaviorally, they avoid temperature extremes when possible. For example, many mammals seek shade when hot. When warm they increase blood flow to the skin releasing body heat to the environment. If the body temperature continues to increase, many animals sweat or pant and the resulting evaporation of water cools the animal. Although effective this method is metabolically expensive and requires water to replace that lost to evaporation. When cold, blood flow to the periphery is reduced thereby reducing loss of body heat to the environment. The skin then acts as a thermal insulator. If the body temperature continues to drop, shivering may result. These involuntary rhythmic contractions generate heat and assure the maintenance of a high and constant body temperature. Additional heat is produced by non-shivering thermogenesis when cold. The cost of such thermoregulation is high and most animals rely on behavior to control their body temperature.

In sharp contrast, cold blooded or ectothermic animals obtain body heat largely from the environment. If one places a lizard, salamander or fish in a refrigerator overnight, the body temperature approaches that of the refrigerator. Obviously such animals do not normally live in a refrigerator. In their natural habitat they often have a mosaic of environmental temperatures available. When cold, many ectothermic animals such as lizards bask in the sun to increase their body temperatures. Many align their body's perpendicular to the sun, maximizing exposure and heat absorption. When hot, they seek shade or orient directly into the sun reducing the surface area exposed to sunlight. By this simple behavioral thermoregulation many ectothermic animals maintain relatively high and stable body temperatures at much less energy cost than their endothermic counterparts.

Animal thermoregulation is complex and the boundaries separating the two kinds of thermoregulation are sometimes unclear. For example hibernating endothermic mammals become ectothermic during hibernation and allow their body temperature to drop in order to save energy during the cold winter months. When food is scarce hummingbirds reduce their body temperature or "hibernate" each night to conserve energy. The swimming muscles of some large ectothermic fish, such as tuna, are essentially endothermic, while the temperature of the rest of the body approximates that of the environment. Some large female snakes generate body heat while incubating their eggs. Even the flight muscles of bumblebees and certain other insects are regulated far above the environmental temperatures enabling more efficient contraction. Other complex intergrades have been described.

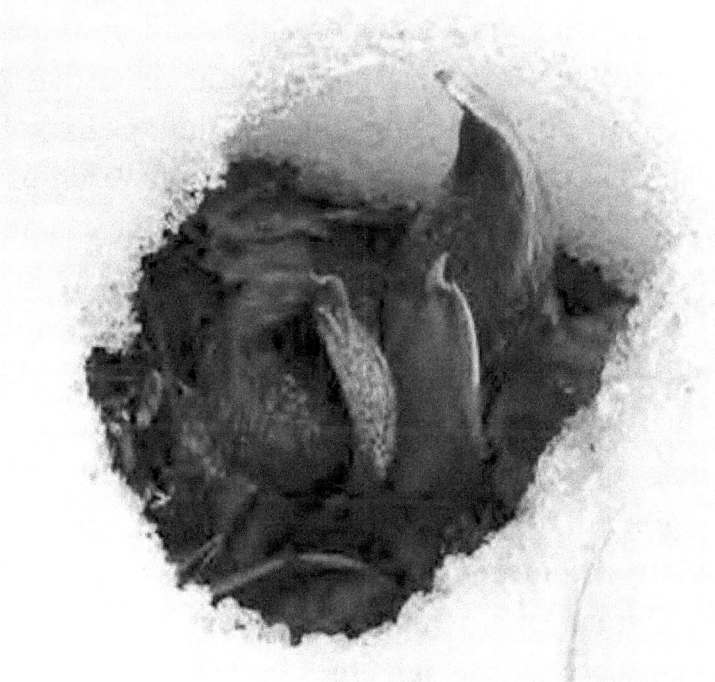

Eastern skunk cabbage, *Symplocarpus foetidus* melting snow
Photo copyright Walter Muma used with permission from:
http://ontariowildflowers.com/main/species.php?id=105

The skunk cabbage is one of those complex intergrades regarding the thermoregulation. Most of the time, its temperature approximates that of the environment as in other plants. In early spring it can however become endothermic and generate enough heat to avoid freezing. The tightly packed club-like shoot or spadix emerges through the soil very early when freezing temperatures often occur. Unlike other cold-hardy plants it is not freeze hardened. The tender spadix of skunk cabbage can not withstand freezing temperatures, yet it emerges early in spring as temperatures still drop below freezing and snow remains on the ground. It has the ability to elevate the temperature of the spadix. For example, when the environmental temperature drops to freezing or 0 degrees C the spadix temperature can at 25 degrees C or 77 degrees F. (Knutson, 1972). The lower the environmental temperature the greater the amount of metabolic heat produced. Most astounding is the actual rate of metabolism of this plant approaches that of a tiny shrew or even that of the flight muscles of a hummingbird during hovering flight. In other words, this plant is physiologically similar to endothermic animals. As the environmental temperatures drop the tiny spadix actually generates enough metabolic heat to avoid freezing. It literally melts the snow around it by heat generated from stored food in the rootstock. This metabolic heating is accomplished by using vast stores of energy from the huge underground root stock. It is difficult to imagine how such a sophisticated and metabolically costly physiological trait; reminiscent to non-shivering thermogenesis found in endothermic vertebrates could have arisen by random variation in the skunk cabbage.

The high rate of metabolism of small birds and mammals has been long known and is well understood. The observation that an isolated group of plants can achieve the same extraordinarily high rate of metabolism as mammals and birds is not only remarkable, but is truly a nightmare for evolutionists. In order to accomplish this amazing feat many things must be available at the same time. For example, the plant must have a huge store of accessible energy without which such heating would be deplete the reserve and kill the plant. It must have temperature sensors to detect near freezing temperatures and a complete system to reverse the normal reduction of metabolism as the temperature decreases. Evolutionists have no tenable explanation for these remarkable physiological traits. *You are worthy, our Lord and God, to receive glory and honor and power, for you created all things, and by your will they were created and have their being.* (Rev 4:11, NIV)

References:
Craighead, J. J., F. C. Craighead, Jr. and R. J. Davis. 1963. Peterson Field Guide Series, Rocky Mountains Wildflowers. PP 11-13. Houghton Mifflin Company. Boston, MA.

Ito, T and K. Ito. 2005. Nonlinear dynamics of homeothermic temperature control in skunk cabbage, Symplocarpus foetidus. Physiological. Review. E72: 051909.

Kirk, D. R. 1975. Wild Edible Plants of Western North America. P. 215. California Naturegraph Publisher, Happy Camp, CA.

Knutson, R. M. 1972. Temperature Measurements of the Spadix of Symplocarpus foetidus (L.) Nutt, American Midland Naturalist 88:251-254.

Pojar, J., A. MacKinnon. 1994. Plants of the Pacific Northwest Coast. Lone Pine Publishing, Vancouver, British Columbia, Canada.

Seymour, R. S., M. C. Barnhart, and G. A. Bartholomew. 1983. Respiration and heat production by the inflorescence of Philodendron selloum Koch. Planta 157(4): 336-343.

Seymour, R. S. and P. Schultze-Motel. 1996. Thermoregulating lotus flowers. Nature 383:305

Smith, E. N. 2007. Endothermic Skunk Cabbage. *Creation Research Society Quarterly*, Fall 2007 Vol. 44(2):153-155.

Bumblebees hum while they work

Bumblebee buzz pollinating a flower of Orphium frutescens
Photo by S. Johnson.
http://www.plantzafrica.com/veldflora/1992/buzzpoll1.htm

Do you sing while you work? This may seem a trivial question, yet in many cultures singing or humming indicates pleasure. It is akin to the purring of a contented cat. People enjoy bird songs as audible signs of joy, yet evolution has forced acceptance of such melodious notes as merely the sounds of one defending its territory or attempting to attract a mate. Several recent scientific studies reveal another animal sings, or at least hums while it works. Its humming has a most unexpected result and presents evolutionists with yet another insurmountable problem.

Many insects and certain humming birds are attracted to flowers as a source of nectar. This sweet fragrant liquid is produced by plants to entice pollinators. Nectar is largely sugar and is an important source of quick energy. Pollen is also produced by flowers to reward pollination. It is up to 50% protein and is a required staple food for endless varieties of adult beetles and bees and an essential food for bee larva.

It seems there is a real buzz (no pun intended) among those who study bees and pollination. Approximately 8% of flowering plants have their pollen tightly locked away from most pollinators, yet must rely on them for cross pollination. For many flowers only a loud sound of a certain frequency will release a shower of falling pollen upon anyone singing the right note. Buzz pollination has been found to be important in the successful cross pollination of flowers from South Africa (Johnston, 1992) to Cranberry bogs of Wisconsin Minnesota and the Pacific Northwest (MacKenzie, *et al.*, 1993 and Heimstra, 1993). Much of the early work in buzz pollination was described two decades ago (Buchmann, 1983). An enterprising university lecturer in Australia even offers to direct field studies of buzz pollination for graduate students. (http://www.anu.edu.au/BoZo/Saul/studentprojects.html)

Certain flowers in the beautiful meadows of

Virginia in the United States change color to attract buzz pollinators at just the optimal time to collect pollen (Milius, 1999). Pollen release is accomplished by a kind of isometric exercise in which the buzz pollinators clasp the flower with their legs and mandibles. By using flight muscles they create a loud buzzing sound, without noticeable wing movement. The sound releases a cloud of pollen quite visible to the onlooker. The buzz is distinct in pitch and noticeably louder than the normal buzzing sound associated with bee flight. To some, the sound is a comical, almost rude sound, akin to one teenager giving another "the raspberry" or "Bronx Cheer". Ah, music to one is mere noise to another.

For certain species a specific tone or frequency is required to release the pollen. A recent National Public Radio airing of **Living on Earth** included an interesting piece about the role of native North American bees in the pollination of important food plants. Dr Sarah Smith Greenleaf, a professor from Princeton University discussed the importance of buzz pollination in tomatoes. According to her, our native bumblebees or one other native bee are required for proper tomato pollination. European honeybees are ineffective because they do not hum while they work.

Tomato pollen is locked inside the flower and is released only when a sound of middle C (261.63 Hz) is present. Honeybees work in silence and thus no pollen is released and they collect scattered pollen left behind by the most recent buzz pollinator. Bumblebees, in contrast hum while they work at exactly the proper frequency to release the tomato pollen. She then demonstrated the process by striking middle C tuning fork and placing it near the tomato flower. A small cloud of yellow pollen suddenly appeared.

There is another even more important reason bumblebees hum while they work. The humming elevates the tempeature of their flight muscles needed to develop enough power for flight. It is interesting that when the science of aerodynamics was new calculations were made about bumblebee flight and it was calculated they could not develop enough power for flight. The problem was the engineers assumed the flight muscles were at room tempearture and if they were the case indeed bumblebees could not fly. Even with an air temperature as low as 6 degrees C or 43 F bumblebees can fly. By simply buzzing their wings, they can elevate the temperature of their flight muscles to 30 degrees C or 86 degrees F enabling them to fly even when the air temperature is cold.

Now you have the pieces. Figure the odds for a bumblebee buzzing in order to elevate its wing temperature so it can fly even on cold days...AND certain plants needing exactly the buzz frequencey in order to release its pollen. At times I almost feel sorry for the evolutionists. This is one of those times. Think of the problems they face! It is they who most hold on to dogma by blind faith. Keep in mind that for any complex structure or behavior to be made by evolution two obstacles must be overcome simultaneously. First, due to the conservative nature of genetic transmission, any "improvement" must occur in small steps. Large, sudden genetic changes are strictly prohibited. Secondly, each tiny step must be advantageous or it will be quickly and irreversibly eliminated from the gene pool by natural selection. What possible advantage could there by in locking pollen away from pollinators? Conversely, why would a bumblebee just happen to expend energy by buzzing their wings while in order to maintain the proper muscle temperature at exactly the frequency required for plants to release pollen? Natural selection would indeed eliminate either response alone. Yet, the opposite occurs. Bumblebees buzz at exactly the proper frequency and intensity to unlock the pollen for pollination. How can any rational mind NOT see design in this? Where there is design there must be a Designer. ALL aspects of this complex symbiotic relation must be complete and functioning before any advantage is possible. Indeed it is the evolutionists that require the greatest portion of faith.

Certainly after the discovery evolutionists argue buzz pollination is important to avoid self pollination and to assure the advantage of increased genetic variability that can only occur with cross pollination. Such a response is neither predictive nor profound. It is a mere afterthought not unlike making the rather obvious conclusion that the barn door was left open after seeing the horses running free. It does not ask the more crucial question as to why the barn door was left open in the first place. So it is with creating "excuses" for buzz pollination. If such a specialized and highly orchestrated event is needed to assure cross pollination then why do 92% of the rest of the flowers freely give pollen to all visitors? Also this accomplishes nothing in explaining how such a unique complex example of symbiosis could develop in small steps from random changes in the first place. If evolution is supposed to provide answers to these and other deep biological questions it fails miserably. Such circular pointless reasoning abounds in evolution and is necessary to prop up a failed idea.

These profound observations demand a Designer and I know the Designer of flowers and bees. Remember there is strong evidence of design in nature and on this we have the Word of God and the facts of science. *Consider how the lilies grow. They do not labor or spin. Yet I tell you, not even Solomon in all his splendor was dressed like one of these.* (Luke 12:27, NIV) *You are worthy, our Lord and God, to receive glory and honor and power, for you created all things, and by your will they were created and have their being.* (Rev 4:11, NIV)

References:

Buchmann, S.L. (1983) Buzz pollination in angiosperms. In *Handbook of Experimental Pollination Biology.* ed. Jones, C.E. and Little, R.J. Sand Editions, New York.

Heimstra, C. (1993) Pollination of cranberries with bumblebees. *Cranberries* 57(3):10.

Johnson, S., (1992) Buzz Pollination of *Orpheum frutescens*. Veld and Flora Articles, *The Journal of the Botanical Society of South Africa*, June, page 36

MacKenzie, K., Cane, J.H., and D. Schiffauer (1993) Foraging by bee pollinators of cranberry. *Cranberries* 57(6): 10, 21-22.

Milinus, S. (1999) Pollination of *Rhexia niginica*, a wildflower. *Science News* April 3.

Smith, E. N. (2007) Do you sing while you work? *Creation Research Society Quarterly* vol 43:261

Also see:
http://www.bumblebee.org/behaviour.htm
http://en.wikipedia.org/wiki/Bumblebee

122

Lunar pregnancy test

Full moon
Wikepedia.org

Long before the corner drug store had inexpensive pregnancy test kits available and before the rabbit test was widely used, God provided women with an easy-to-use absolutely free pregnancy test. I shared this with my nursing students and was shocked that not a single female student was aware of this simple test. As a young man, when I first learned about a woman's reproduction cycle, I instantly put it together with my knowledge of the lunar cycle and was in awe of the Creator for making such important information readily available to all women long before calenders were available.

In simpler times before the Internet, cell phones, television and electric lights, people were closely tied to natural light and the phases of the moon. God's Word clearly states: ***And God said, Let there be lights in the expanse of the sky to separate the day from the night, and let them <u>serve as signs</u> to mark seasons and days and years*** (Gen 1:14, NIV). Early civilizations were aware of this and marked such celestial events as spring and fall equinox and the longest and shortest days. Long before the advent of months, years, and even before calendars were thought of, a woman could simply go outside at night, look at the moon and determine if she might be pregnant.

How could they do this? It is really quite simple. God designed the lunar cycle to be roughly twenty-nine days in length. This is the period from new moon to new moon or full moon to full moon and is amazingly close to the average woman's 28 day menstrual cycle. Here is how it works. If a woman started her period last month with a full moon and the next full moon came without her having a period she knew she immediately that she might pregnant. How could it be simpler? Certainly this is not 100 percent accurate, for the menstrual cycle can vary, but it still provided women with an early indication of possible pregnancy.

This raises yet another problem for evolutionists. Why would the lunar cycle agree so closely with a woman's menstrual cycle? Did people evolve so the moon could be used as a pregnancy test? Not likely! Or did the two events just happen to coincide? Again most unlikely; figure the odds! It is far more logical that the God who hung the moon and stars in place and created man in His image planned the cycle of the moon so it could be used as a "sign" of pregnancy. Today we think of the moon as being for lovers. So did our Creator. The phase of moon has been used since the beginning of Creation to tell a woman if she was with child. *The heavens declare the glory of God; the skies proclaim the work of his hands* (Ps 19:1). Sadly, such truths can no longer be taught in our public schools or universities. Our teachers are strictly forbidden to mention any evidence from science that points to Intelligent Design, creation or to God. Instead, they are required to teach materialistic evolution and that man is nothing more than "matter in motion." This must change. Teachers, like scientists should be allowed to follow the evidence, no matter where it leads and a great deal of scientific evidence leads directly to the God of Creation.

Reference
For additional information about the lunar phases visit this excellent website.
http://facstaff.gpc.edu/~pgore/astronomy/astr101/moonp has.htm

Cicada killer, neurosurgeon extraordinaire

© Z. Huang

Yellow and Black Cicada Killer with prey
Photo by Zachary Huang, www.cyberbee.net

The beautiful yellow and black cicada killer, *Sphecius speciosus* is the largest North American wasps. As a boy, I watched many times as they captured and carefully sedated cicadas (called "locusts" in Oklahoma) which are about three times their size. First, they would systematically sting the nerves that control the flight muscles, rapidly subduing the huge insect. Next, one by one, they would sting the nerve ganglia associated with movement of each leg. Once the prey was effectively paralyzed, she would climb the nearest tree trunk or fence post dragging the huge cicada along behind. She would then fly off, clutching it tightly with their legs. The weight was too much and she was unable to maintain altitude and would gradually sink to the ground.

Upon landing, she would again drag the heavy prey up a nearby post or tree trunk and repeat the process until she finally returned to the previously dug burrow. She would then drag the prey underground and lay a single egg in it. The entire process would take an hour or longer. Years later, I found out the cicada must remain paralyzed, but alive to nourish the cicada larva. The larva instinctively eats around the vital organs until it is nearly mature. Finally, it eats the vital organs, killing the cicada and pupates. Even as a young boy, I was impressed at the instinctive knowledge the adult cicada wasp had of the cicada's nervous system as well as its ability to find the burrow she had dug. Stop and ponder the complexity and evidence of design here. It is utterly impossible a female wasp could accidently find each of the nerve gangula that control movement in the giant cicdia...or know how much venom to inject. To much would kill the prey and her young would starve...to little and it would fight the young again causing it to starve.

It bogles my mind that some actually believe all this happened by accident in many tiny steps by trial and error of vast periods of time. Remember each step MUST be advantagenous or it will be eliminated by natural selection. If you believe that...might I sell you some beach front property in Arizona? The evidence of design is overwhelming in this example alone. What am

I missing? How can anyone not believe? The unmistakable evidence of God is all around us. To me as a boy and especially now as a scientist the evidence of God as Creator is overwhelming. I find it strange with all we know about the complex structure and behavior today that anyone can still cling to the belief that all this just happened without direction in small inheritable steps by random mistakes in the genetic code. Truly, as God's Word declares only the fool fails to see the hand of God in nature.

References

http://entomology.ifas.ufl.edu/creatures/beneficial/cicada_killers.htm

Dambach CA, Good E. 1943. Life history and habits of the cicada killer (*Sphecius speciosus*) in Ohio. Ohio Journal of Science 43: 32-41.

Holliday CW. (2009). Prof. Chuck Holliday's Cicada Killer Page. Lafayette College. http://ww2.lafayette.edu/~hollidac/cicadakillerhome.html (20 July 2009).

Evans HE, O'Neill KM. 2007. The Sand wasps: Natural History and Behavior. Harvard, Cambridge, Mass. 340 pp.

Lin N. 1963. Territorial behaviour in the cicada killer wasp, *Sphecius speciosus* (Drury). I. Behaviour 20: 115-133.

Temperature determines alligator sex

There are many difficult problems for evolutionists to explain that few people know about. Because they present insurmountable problems for evolution, they are not mentioned in textbooks or discussed in biology classrooms. One of these difficult problems is related to the determination of sex in alligators and turtles. First, some background will be helpful. Remember, I studied these fascinating creatures for most of my life.

Female alligator defending her nest
Photo by the author.

Alligators must be eight to ten years old and approximately six feet long to breed. Most mature females breed three out of every four years. They construct a large nest near the water from nearby vegetation. When complete they deposit an average of forty-seven eggs in a depression near the top of the nest. After the eggs are laid, she covers them with more vegetation. Alligator eggs require nine to twelve weeks to hatch and are incubated by the heat from the sun and heat produced by of the decaying vegetation. During the time the eggs are incubating, the female alligator has a restricted home range of less than a third of an acre and will actively defend the nest. Just her presence deters most natural enemies.

Two or three days before hatching, baby alligators inside the egg begin making a grunting sound which synchronizes the hatching so all the young alligators emerge from the eggs within a few hours. Upon hearing the sound of her emerging young, the mother alligator carefully removes the nest material from the eggs to release the hatchlings. If the female has been killed, the young alligators remain imprisoned inside the nest and die. Mother alligators have been seen gently breaking the egg shell with their teeth when a baby alligator has difficulty hatching. The female will take each baby alligator into her mouth and carry them to the water where she releases them and returns to the nest for another one. In areas where alligators winter in underground burrows the young stay with their mother the first two years. Such maternal care is uncommon among reptiles and is largely unreported by evolutionists.

They teach students that maternal instincts did not evolve until the appearance of birds. God was not restricted by evolution dogma and made all living things optimally adapted for their environment. ***God saw all that he had made, and it was very good*** (Gen 1:31a).

Alligator eggs in the nest
Photo by the author

It is the incubation temperature and not sex chromosomes that determine the sex of baby alligators. This presents an even more difficult problem for evolutionists. Alligator reproduction is well known and often shown on television nature programs. Alligators and certain other reptiles lack sex chromosomes (Ferguson and Joanen, 1982). Ted Joanen and Larry McNease of the Rockefeller Wildlife Refuge in Grand Chenier, Louisiana taught me how to call, capture and safely handle alligators and I have returned to the refuge several times. Dr. Ruth Elsie and others continue alligator research at the refuge. The refuge has a higher concentration of alligators than any other place in the world and is a beautiful place to visit, watch birds and observe alligators in their natural environment.

Research has shown that an average incubation temperature of 85 degrees Fahrenheit yields only female hatchlings. At a temperature of 89 degrees, equal numbers of male and female hatchlings result. At 91 degrees, only males are produced. The investigators admitted in the original paper that, "There has been no demonstration of a selective evolutionary advantage of the occurrence of temperature sex determination in reptiles." One is hard pressed to present an argument in favor of this sort of sex determination as opposed to the more common method of sex chromosomes. If there is a selective advantage, then why do most animals rely on sex chromosomes?

These are very difficult problems for evolutionists, but there is an even worse problem. If one could conjure up a selective advantage for such results in crocodilians, the argument runs amuck with the consideration of turtles. Most turtles also lack sex chromosomes and the sex of the offspring is determined by the average incubation temperature of the eggs, but there is a remarkable difference. In turtles eggs incubated at higher temperatures produce females, not males! The notable exception is *Trionyx spiniferus* the beautiful and common spiny soft-shelled turtle which have sex chromosomes and the sex of the hatchlings is unaffected by egg incubation temperature (Bull and Vogt, 1979).

This leaves the evolutionist with an unenviable situation. Whatever arguments are fabricated to support temperature induced sex determination in crocodilians disintegrate when one considers turtles. It is not surprising evolutionists are unwilling to address these important problems for they know they are standing on shifting sand. It seems logical that accepting any of these "excuses" evolutions might imagine for such sex determination is foolish. It is difficult to imagine some yet-to-be-conceived alleged advantage of allowing temperature to determine sex one way in crocodilians and the opposite way in most, but not all turtles. Once again it is not a good time to be an evolutionist. Facts like these continue unrelenting to give them nightmares. Perhaps there is another reason for this unusual mode of sex determination.

Could this whole sex-in-response-to-temperature thing have been designed by a Creator who likes variety? Or perhaps He is simply showing a sense of humor. Yes, God DOES have a sense of humor. For proof consider the account found in the eighth chapter of Exodus. One of the things God used to reveil his power to the Pharaoh was the plague of frogs. That is a well known passage, but many people miss the humor because they do not know the Egyptians worshiped a frog god. God was saying, "You like frogs, I will send you frogs."

References

Bull, J. J. and R.C. Vogt. 1979. Temperature-Dependent Sex Determination in Turtles. *Science* 206:1186-1188.

Ferguson, M. W. J. and T. Joanen. (1982). Temperature of egg incubation determines sex in Alligator mississippiensis. *Nature* 296, 850-853.

Kangaroo rats conserve water

Ord's kangaroo rat, *Dipodomys ordii*
(www.biotopics.co.uk)

While walking one morning at sunrise in the beautiful desert of southern Arizona, I realized how truly underrated kangaroo rats are and what a problem they are for evolutionists. They occur on my farm in western Oklahoma, but here in the desert they were abundant with burrows and fresh tracks everywhere. Kangaroo rats are beautiful little brown and white animals with huge liquid brown eyes and a perky gait akin to kangaroos. As we will see, they have given us so much, yet we have returned little. Let me explain.

One of their remarkable features is the number of facial muscles, yet this is never discussed in biology classrooms, partly because if flies in the face of established evolutionary thinking. During the twentieth century it was widely publicized that of all mammals, primates had the most facial muscles. The large number of facial muscles is necessary for expression and is used in non-verbal communication. This remains widely accepted, in spite of well known fact that an elephant's trunk possesses far more with over a thousand individual muscles. Primates do in fact have over twice the number of facial muscles of most other mammals. Emotions of all primates including man can be read by changing facial expressions. We see no corresponding facial grimaces in other animals such as livestock, dogs or cats. It was concluded that such communication was correlated with social behavior and intelligence and thus the maximum development and diversity of facial muscles were seen in the most intelligent animals. Man is always seen as the glorious climax of evolution. Such simplistic thinking seldom withstands the test of time.

Remember, this is the same crowd that for decades taught the human brain was the largest and most highly evolved of all animal brains. This was until scientists noticed the brains of whales and even the pelvic brain of certain dinosaurs (an enlargement that controls the hind legs) was larger than human brains. Evolutionists quickly modified the definition to say the human has the largest brain as percentage of body mass and thus is proof man is the most highly evolved of all creatures. Once again those pesky facts keep getting in the way of good evolution dogma. Other animals were soon found to have larger brains even as a percent of body mass. The theory was once again modified to state the human brain is the most convoluted and thus has the greatest surface area of any animal. Evolutionists have not yet recovered from the discovery that whales and porpoises have both relatively larger and more convoluted brains than humans. Again, those nasty facts keep getting in the way of good evolution legend.

Let's return to our discussion of kangaroo rats. Several years ago, a graduate student discovered kangaroo rats have a surprising forty-two pairs of facial muscles! So much for the theory that intelligence is correlated with the number of facial muscles. Of what possible reason could a small rodent use with so many facial muscles? Such an obvious question demands a thoughtful answer. Even the casual observer realizes kangaroo rats do not appear to spend much time making faces at one another.

Here the unique kangaroo rat's gait comes into play. They are often bipedal...hopping like their larger Australian namesakes. But they do not hop very high. Night photography revealed a striking discovery. As they hop, their vibrissa or sensitive "whiskers" maintain contact with the surface! Their need for such a large number facial muscles is for the purpose of providing detailed tactile information about their environment as they hop about. Of course, nothing is written about how such a complex dynamic surface scanning system and its extraordinarily complex interpretation might have evolved. The insignificant kangaroo rat puts yet another fly in the evolutionist's ointment, but there is more...much more.

Kangaroo rats make excellent pets. I kept several as a kid growing up on a farm. This was before they were protected by law and it was legal to keep such creatures. They require little care and are delightful to watch. They eat little and require absolutely no drinking water. Unlike most other mammals, they are virtually odor free. Their lack of any need for drinking water is perhaps their most outstanding physiological feature. For years it was observed that many desert inhabitants require little or no water. Scientists have worked out daily water budgets for a wide variety of animals including desert creatures. They carefully measured all water intakes and collected urine and feces to determine the amount of water lost. Respiratory and skin losses of water were also measured. Perplexed by what they found, some biologists postulated dew as a source of water. Obviously those biologists had not spent much time in the desert for dew is a rarity and any animal depending on dew for hydration would not long survive.

Other biologists suggested the importance of "pre-formed" water found naturally in the desert. Even certain desert plants contain high concentrations of water.

Careful observation however clearly showed kangaroo rats could live without such plants. They can subsist entirely on a diet of dry seeds. This means they thrive on the metabolic water produced when carbohydrates are used producing carbon dioxide and water. We all remember from grade school that when sugar is metabolized the end products are carbon dioxide and water. This is the only water required by these delightful little creatures.

Examination of their urine was another biological bombshell. They can produce urine three times more salty than seawater! This observation eventually led to a better understanding of how humans produce urine. The most important functional part of the kidney is the loop of Henley. This is the portion that concentrates urine and saves water. It was discovered the loop of Henley of kangaroo rats was proportionally three times longer than it is in humans. Farther studies showed the primary function of the loop of Henley to be water absorption and retention. Thus an understanding kidney function in kangaroo rats led directly to a better understanding of how the human loop of Henley functions in humans.

Such is often the case. The human body can do many things (physiologically speaking) but cannot do any of them particularly well. It is when we study extreme animal forms, those living on the physiological edge so to speak, that we better understand human function. Rattlesnakes can live up to two years without food or water and parasitic ticks can survive five years with neither blood nor water! No doubt they, too, have secrets we will someday unlock. See why I love zoology? See how our Creator reveals Himself in nature?

It is because the kangaroo rat can produce urine three times more salty that they can survive in the desert southwest without the need to drink water. Humans can only produce urine slightly more salty than seawater. We have all seen movies of those lost at sea and some might wonder then why we cannot survive by drinking seawater. The answer is that the third most abundant salt in seawater is magnesium, which causes excessive water loss from diarrhea when sea water is ingested in quantity. We have learned much from the kangaroo rat and no doubt more remains to be learned. What do you think about while walking in the desert? I think of kangaroo rats and their all-wise Creator and find myself lifting my hands in joyous praise.

References

Hoditschek, Barbara and Troy L. Best, 1983,

Reproductive Biology of Ord's Kangaroo Rat (Dipodomys Ordii) in Oklahoma. J. Mamm. 64(1):121-127.

Ord's Kangaroo Rat, Wikepedia online encyclopedia

Ichneumon wasps

Ichneumon wasp laying its egg
Photo by Gene Hicks

This beautiful wasp has many features that display the Creator's wisdom, but present a plethora of implausible obstacles for diehard evolutionists. These insects appear more fly-like than wasp-like and are sometimes incorrectly called "Ichneumon flies." Due to their greatly elongated curved abdomen they are also wrongly called "scorpion wasps." There are roughly

60,000 species of ichneumon wasps worldwide and 3,000 species in North America, making it the most abundant member of the Hymenoptera order of insects. They are the most notable exception to the commonly accepted latitudinal gradient rule because, unlike most other species there are more ichneumons in higher latitudes then closer to the equator. No doubt this exception occurs because of most ichneumons require host insects for their young that feed on hardwood trees.

Ichneumon wasps do not sting as a defensive weapon. The females possess an ovipositor that is longer than her body. Some females inject venom into the prey along with the egg. The venom paralyzes but does not kill the prey larva for their young. Males lack the ovipositor. Both males and females are often found on tree trunks rhythmically tapping on the wood with their antennae. Males do this in an attempt to find emerging females with which to mate. Females do this in order to find wood boring larvae in which to deposit their eggs.

Once the female finds a prey species she then bores the ovipositor through the wood and deposits the egg in the prey insect. It remains unknown how the soft, flexible ovipositor can actually bore through hard wood to reach the larva. Recent research has revealed metal in the form of ionized manganese and zinc at the extreme tip of the ovipositors of some ichneumon's ovipositors. Such a hardened tool is not found in another other animal. Of course there is an even more difficult problem…how can she possible know from the outside of a tree trunk precisely where such a larva is by merely tapping on the wood?

Ichneumon wasps truly present multiple nightmares for evolutionists I have long found it interesting that such problems for evolutionists are ignored in biology textbooks and lectures by professors. They are simply swept under the carpet and remain largely unknown to most students. The ichneumon wasp is an excellent example. Always keep in mind evolution has two powerful restraints. Each step in the evolution of a structure, molecule or behavior must be adaptive...it must offer some selective advantage or it will be forever lost by natural selection. This alone has shaken the very foundation of evolution dogma today with each new discovery of the complexity of subcellular tools such as the flagellum. There is an additional caveat even more difficult to overcome. To influence the gene pool each new trait, behavior or structure must be inherited, but such changes can not be very large, due to the conservative nature of genetics. Lets look closely at some of the unique features of ichneumon wasps.

Again, they have many unusual behavioral traits and structures that make them well suited for their ecological niche. Consider the size and strength of the female's ovipositor. It is longer than its own body and the length is necessary for depositing eggs inside a living beetle larva deep inside a hardwood tree. From what did it evolve? A half developed ovipositor would be eliminated from the gene pool. How did the end of the ovipositor become hard enough to penetrate over a centimeter of hard wood? How did it "discover" how to collect metals at the end for strength? How did it learn to find and recognize a suitable larva for its young by tapping on the tree with its feet? How did it develop the proper amount of venom to paralyze, but not kill the larva food for its young? According to evolution such incomplete or imperfect intermedate steps would NOT have a selective advantage and would be quickly and forever eliminated. Evolution totally fails to even address, much less come up with rational explanations for these and many other questions.

Besides all the unique and highly specialized anatomical features consider some of this wasp's unique behavior. Keep in mind learned behavior can not be passed on the young. The idea was proposed in ancient times by Hippocrates and Aristotle. It was also supported by Darwin. The French naturalist Lamarck greatly expanded on this concept as a central mechanism by which evolution operated. As our understanding of genetics advanced Lamarckism was rejected because there is no mechanism for passing on acquired traits of the parents to the offspring. Perhaps the strongest proof of the fallacy of such a view is the observation then Jewish boys have had no reduction in their foreskin after thousands of years of circumcision. Again, learned behavior can not be passed on to the offspring. Such behavior it must be both instintive and advantagous.

146

The ichneumon wasps have a multitude of complex unique behavior patterns. How could the male acquire the behavior trait of tapping on a tree in order to discover when a new young female was about to emerge? Even if it was lucky and made such a discovery there is no mechanism by which it could be passed on the its male offspring. The same is true for the female tapping on the tree truck to find a wood bore larva deep inside the tree. Blind luck can simply not explain how such a complex and important behavioral trait could develop in the adult in the first place and certainly can not account for how this is an inherited trait. Many other examples could be listed. Once again evolution fails totally to account for the complexity we find everywhere around us in God's Creation. Again it is the evolutionist that has the greater faith…for I KNOW in Whom I have believed. Truly, ***Through him all things were made; without him nothing was made that has been made.*** (John 1:3, NIV)

References:

Ichneumonidae, Wikepedia online encyclopedia

Jackman, John A. and Bastiaan M. Drees. 1998. A field guide to common Texas insects. Taylor Trade Publishing, USA.

Giraffe blood pressure

Giraffe eating

Giraffe drinking

Chaffee zoological Gardens of Fresno, CA

I have long thought zoologists are born, not made. Educaton only refines the interests that are already present. Let me give two personal examples. Buchering day was a big event on the farm where I grew up. Several family members were present in order to butcher, cut up and can the meat. This was before we had electricity and freezers. Even as a young boy, I was interested in the internal organs of the steers, hogs and even chickens that we butchered. I had two burning questions about each inernal organ. What is this? And what does it do? Those same two questions have driven much of my research as a zoologist.

When most people see a giraffe for the first time, they are amazed at their height and indeed they are very tall. They are by far the tallest of any animal living today and can get up to 20 feet (or 6 meters) tall. Ever the zoologist, I remember upon seeing my first giraffe I asked my beloved Grandfather, "How does its heart get blood way up there to its head?" That question has been on my mind for decades and I finally have some answers.

Their heart must pump blood up 3.5 meters (or 11.5 feet) to the head while eating and while drinking the head is 2.5 meters (or 8.2 feet) below the heart. As expected their hearts are huge and can weight up to 26 pounds (12 KG). The viscosity of their blood is similar to that of humans. Actually, the answer to how they get blood to the brain is obvious…they have much higher blood pressure than any other animal. Their measured blood pressure is typically 240/160 mm Hg compared to the human's 120/80.

OK, that solves one problem and explains how the blood gets to their brains when they are standing and feeding on folage high above the ground. The even more difficult issue is why the don't have a stroke when they bend down to drink water. Surprisingly, measurements showed that the blood pressure at the base of the brain was 200 mm HG (millimeters of Mercury) when the giraffe is upright and, instead of being higher as expected, dropped to 175 mm Hg when the head was lowered.

As in most ruminants, the blood reaches the brain from the heart by way of the common and external carotid arteries. The two external carotids divide, just before reaching the brain, into many small vessels forming a tight network that is called the rete mirabile. The vessels of the giraffe rete have elastic walls which can accommodate excess blood when the head is lowered so that the brain is not flooded. As a further safeguard for the brain while the giraffe is in this position, a connection between the carotid artery and the vertebral artery drains off a portion of the blood even before it reaches this network. The walls of the rete mirabile vessels are also elastic enough to retain sufficient blood when the head is raised so that the brain's supply is not depleted momentarily during the system's pressure changes.

In other words the giraffe has a complex and well designed mechanism and anatomical features to assure blood pressure is maintained at normal levels whether the animal is feeding on leaves high up in trees or drinking water from below the level of its feet. That such a complex system could develop by small mutational errors is ridiculous beyond comprehension. Once again the more we learn about the complexity of ALL living things…the stronger the evidence they were designed by an all knowing Creator.

References

Dagg, Anne Innis and J. Bristol Foster, 1976 *The Giraffe, Its Biology, Behavior, and Ecology*, Robert E. Krieger, Publishing

Giraffe, Wikipedia, the free encyclopedia

http://wiki.answers.com/Q/What_is_a_giraffes_normal_blood_pressure#ixzz1KeBKtGKR

Issues of the Heart
An organ of exquisite design
Nature uses as little as possible of anything. Johann Kepler

When one looks closely at the hearts of various vertebrates it is obvious each possesses an ideal heart of exquisite design, yet for decades their hearts have been used as evidence of evolution. As a physiologist, I find it interesting that a lack of understanding leads to erroneous assumptions about evolution. This happened early in the history of evolution when Darwin's staunch defender, Ernst Haeckel prepared a list of 126 organs in the human body that he considered useless. These vestigial organs were seen as left over relics of our evolutionary past. As knowledge replaced ignorance we discovered we needed such "useless" organs as: pituitary gland, parathyroid glands, tonsils and appendix. I can live without my appendix or my right arm, but that does not mean God did not give them to me for a purpose.

Today no part of the human body is considered vestigial and the argument has lost its effectiveness. Sadly, countless lives were lost when physicians eagerly embraced this erroneous teaching and removed the appendix and other organs thought by evolutionists to be useless. I am reminded of the words of King David: *For you created my inmost being; you knit me together in my mother's womb. I praise you because I am fearfully and wonderfully made; your works are wonderful, I know that full well. My frame was not hidden from you when I was made in the secret place. When I was woven together in the depths of the earth, your eyes saw my unformed body. All the days ordained for me were written in your book before one of them came to be.* (Ps 139:13-16) Ernst Haeckel would have gained more lasting respect had he studied the teachings of the King David rather than Saint Darwin.

It seems a universal human trait to attempt to bring order from disorder. In the Garden of Eden God gave Adam the responsibility of naming the endless array of living things, thus bringing order. More recently the science of taxonomy has become an attempt to arrange various species in phylogenetic order showing their assumed evolutionary relationship. Even if evolution is totally false, taxonomists do a service by bringing order to the myriad of living things past and present. Problems arise when these imagined evolutionary schemes are used as evidence of evolution or to make predictions. This motivation to see phylogenetic progression as evidence of evolution led to decades of misunderstanding of the exquisite design of the heart of many diverse vertebrates. Let's consider some of theses issues of the heart.

Vertebrate circulation

The most significant difference between the cardiovascular system of vertebrates and most other animals is they possess a closed circulatory system. In all vertebrates the blood is contained in vessels in contrast to the open circulatory system of most invertebrates. Understand the cephalopod mollusks such as the giant squid are exceptions and also have a closed circulatory system. This is yet another annoying fly in the evolution ointment and is carefully avoided in evolution lectures and biology textbooks.   Most invertebrate species that possess a heart have an open system in which blood slowly makes its way back to the heart after flowing outside blood vessels and directly around other tissue. Body movement aids in returning blood to the heart of insects, some mollusks and other animals lacking backbones. Even here as mentioned above there are exceptions that present nightmares for evolutionists.

For decades biology students were taught an over simplistic view of the assumed evolution of the cardiovascular systems of vertebrates. It was motivated in part by an attempt to bring order and simplicity from complexity, but primarily to show the phylogenetic progression demanded by evolution. All serious biology students are familiar with a series of simplified drawings appeared in biology and comparative anatomy textbooks showing the "advance" of "higher" vertebrate hearts from the simple two-chambered heart of fish, to the three chambered heart of amphibians, the four chambered but incompletely divided ventricles in non-crocodilian hearts, the complete four chambered but leaking (while diving) heart of crocodilians to the complete four chambered heart of birds and mammals. Much emphasis was placed whether or not oxygenated blood from the lungs or gills mixed with unoxygenated blood returning from the body. As modern blood flow techniques were applied it was obvious many of the original assumptions based on the requirements of evolution and the dissection of dead animals were in error. In spite of new information the oversimplified and erroneous drawings remain as evolution dogma in the textbooks.

Before we compare the hearts and circulatory systems of vertebrates let's review some of the important functions of the cardiovascular system. Certainly gas transport is important, removing carbon dioxide from the tissue and supplying an endless supply of oxygen. Gas is exchanged to the environment in the various vertebrates through the skin, by gills and lungs or by a combination of these. Oxygen requirements are higher for warm blooded birds and mammals. Heat is also transported by the circulatory system and plays a major role in the regulation of body temperature in birds and mammals. The blood transports nutrients from the gut or storage areas such as the liver and adipose tissue to where it is needed. Blood provides the required hydrostatic pressure for proper kidney function. Hormones are transported in the blood from the endocrine glands of origin to the various target organs throughout the body. There are numerous additional blood functions such as the role of certain white blood cells in fighting disease and the complex regulation and control systems to maintain blood pressure, blood volume and the proper numbers of the various kinds of blood cells. Loss of blood evokes a series of life saving reflexes that tend to reduce blood loss and return blood volume and oxygen carrying capacity to normal values quickly.

Gas transport is considered to be the most critical of all cardiovascular functions and indeed drives many of the homoeostatic regulatory mechanisms. It is here evolutionists see a progressive development from fish to amphibian to reptile to bird and mammal. The evolution story is told of the linear progression of ever increasing complexity and efficiency from the simple two chambered heart of fish, to the three chamber "leaking" heart of amphibians, the imperfect heart of reptiles and crocodilians to the perfected and efficient hearts seen in birds and mammals. As is often the case in evolution dogma, the actual facts belie such orderly progression. Clearly intelligent design and not evolution can be seen as we examine vertebrate hearts in light of modern research.

Fish hearts fail to support evolution

At first glance, comparative vertebrate heart anatomy does seem to support fish to mammal progression. As often occurs in science, the devil is in the details, and the details are not revealed to beginning biology students. It is true many fish have a relatively simple two chambered heart with the entire cardiac output going directly to the gills. It should be noted that some authors consider the non-pumping reservoirs (sinus venosus leading to the atrium and bulbus in teleosts or conus arteriosus in elasmobranchs following the ventricle) as an additional two chambers giving fish a four chambered heart. Either way there are only two actually pumping chambers compared to 3 or 4 in other vertebrate hearts. Blood pressure is greatly reduced as it passes through the gill capillaries. From the gills, blood profuses slowly to the rest of the body. There are over 27,000 known species of fish, more than all the other vertebrates combined. With this amount of diversity one would expect many variations regarding respiration and circulation to occur and this is indeed the case.

Hagfish have a series of supplemental hearts aiding in the return of venous blood to the primary heart. Perhaps even more astounding is these ancillary hearts are powered by skeletal muscle and require neural impulses for each heartbeat. In contrast, all other vertebrate hearts are made of cardiac tissue and possess a built-in rhythm or pacemaker. The evolution of these accessory hearts along with their complex neural control or their subsequent fates are never discussed. The reason is simple. Such complexity in such a "lowly" hagfish does not fit the accepted evolution dogma. Lungfish hearts are perhaps the most complex and thus the most interesting of all fish hearts. Anatomically they have the normal two chambers plus some complicated ridges inside the ventricle. For decades the function of these ridges remained unknown and evolutionists saw the lungfish heart as simple and inferior. Since the oxygenated blood from the lungs and unoxygenated blood from the body returned to the same atrium and ventricle it was assumed that inefficient mixing occurred. Such mixing would waste energy, yet such gradual development is required by evolution. Radiographic studies reveal the complex heart of lungfish preferentially pumps unoxygenated blood to the lungs and oxygenated blood from the lungs to the rest of the body. This is accomplished by the remarkably complex series of ventricular endocardial ridges. This efficient mammal-like unmixed circulation in such a "lower" vertebrate flies in the face of the over simplistic fish to mammal progression required by evolutionists, yet it is to be expected with intelligent design. It is also most instructive that fish with only two pumping heart chambers can accomplish what the much larger and more complex four chambered mammalian heart can do. Truly, as Kepler said long ago: "Nature (read: "God") uses as little as possible" to accomplish the goal. More trouble lies ahead for the evolutionist.

Amphibian hearts create havoc

Amphibians, like fish vary greatly as a group and occupy a variety of habitats from desert to fully aquatic life and endless niches in between. Most have well defined right and left atrium with blood from the lungs entering the left atrium and blood from the body going to the right atrium. Some lack lungs entirely and remain their entire lives in water. A few lack a fully divided atrium with fish like hearts. Many venture far from water and rely exclusively on lungs as adults. A unique and prominent spiral valve is present in frogs and may contribute to separating of oxygenated and unoxygenated blood for systemic circulation. Many frogs that normally respire with lungs hibernate underwater and use their skin for gas exchange in winter. Similarly the literature is contradictory with some investigators claiming inefficient mixing of oxygenated and unoxygenated blood in the common ventricle while others see total separation. With such variety and confusion in the literature it is difficult to understand why the simple, but untruthful and misleading picture of amphibian circulation showing progression from fish to reptile remains in textbooks.

Instead of evolutionary progression it is far simpler to see each amphibian heart designed for its way of life. One would expect pulmonary profusion to be reduced or turned off in a hibernating frog which obtains its oxygen from its skin and little mixing of blood if it is relying on its lungs for gas exchange. Perhaps the key is to understand the fluid physics characteristics of the mysterious spiral valve. Here farther research is needed.

Rebellious reptilian hearts

Reptiles, like fish and amphibians are a diverse group. Sea snakes are pelagic living in the open sea.

Other reptiles live in the harshest of deserts while many survive in tropical rain forests or live in freshwater ponds and rivers. There are two contrasting types of reptilian hearts. Non-crocodilian snakes, turtles and lizards have four chambered hearts with an incomplete ventricular septum while crocodilian gavials, caiman, crocodiles and alligators have a completely divided ventricle. The ventricle of non-crocodilian reptiles is divided three distinct sub-chambers, a central canal and a complex muscular ridge. Due to the incompletely divided ventricular septum non-crocodilian reptiles were first thought to have inefficient mixing of oxygenated and unoxygenated blood. Farther studies have shown this not to be the case, but the fluid dynamics are complex and are not fully understood. It is clear reptilian hearts are far more complex and certainly more efficient than was first alleged based on simplistic evolution progression.

Crocodilian hearts are tantalizingly complex

They have essentially the heart of a bird or mammal, with one profound caveat. (For more detail see an earlier section, *Alligators dive for hours*) They have a aperture, the Foramen of Panizzae bypassing the lungs while diving. Again the crocodilian was classified as one mixing oxygenated and unoxygenated blood based purely on the anatomy. As with other vertebrates studied there appears to be absolutely no mixing of the oxygenated and unoxygenated blood. Crocodilians retreat from threat by diving underwater. Such dives may last several hours.

During prolonged diving blood bypasses the lungs by way of the Foramen of Panizzae. Instead of being viewed as an inefficient part reptile-part mammal heart it is now seen as an exquisitely designed organ perfectly fitted for its owner. In fact, most comparative physiologists see it the crocodilian heart far superior to the mammalian heart for its aquatic way of life. A truly four chambered heart would waste energy profusing lungs, yet pick up no oxygen during prolonged diving. Of course the alleged evolution of this complex organ and the required intermediate steps are never discussed in textbooks or classes for their acceptance without any supporting evidence would require a leap of blind faith most students would not make. Again the evidence is for intelligent design and we can know the Designer personally!

Bird and mammal hearts contradict evolution

The evolutionist sees increasing complexity and efficiency as one goes from fish to amphibian to reptile to bird and finally to mammal. Certainly birds and mammals do have the most sophisticated heart of any other vertebrate. There can be no mixing of oxygenated and unoxygenated blood, yet this alone is not the complete picture. Both birds and mammals are warm blooded with a higher rate of metabolism than other animals and changing patterns of blood flow make such high and stable body temperatures possible. Mammals are thought to be more advanced than birds, yet in at least two ways the circulatory system of birds seem superior.

Flying is hard work, much harder than walking or running. Birds often have a higher body temperature and higher rate of metabolism than mammals. Perhaps most surprising is birds have a more sophisticated respiratory system than do any other vertebrate. By a complex series of air sacs scattered throughout their bodies air is passing continuously past the tiny air tubes where gas exchange occurs. This is in contrast with mammals that must inhale, pause, exhale and so forth. There is also a countercurrent system in bird lungs missing in mammals. In bird lungs the blood flows in the opposite direction than air flow maximizing the amount of oxygen that can be transferred to the blood. It is shocking that this well known fact seems missing from the introductory textbooks because it does not fit the assumed evolutionary pattern. Such a countercurrent system of blood flow is also found in the gills of fish, yet this too is often omitted.

Summary

When one looks in some detail at the anatomy and function of the hearts of various vertebrates a clear picture emerges. Each animal has a heart of exquisite design performing its required job efficiently. There is absolutely no orderly progression from simple to complex as one looks in turn at the hearts of fish, amphibians, reptiles, birds and mammals. This is once again seen as strong support of intelligent design. In sharp contrast it presents an endless array of nightmares for evolution with its requirement of progressive development. Perhaps it is time for a paradigm shift. Evolution has failed and is totally bankrupt in explaining the origin and complexity of the hearts of vertebrates.

Often people in the arts see things as they are more clearly than scientists. They can also present such wisdom in ways that bring joy. Let me close with one of my favorite poems. In it Robert Service refers to another organ of exquisite design.

THE WONDERER

I wish that I could understand
The moving marvel of my hand;
I watch my fingers turn and twist,
The supple bending of my wrist,
The dainty touch of fingertip,
The steel intensity of grip;
A tool of exquisite design,
With pride I think: "It's mine! It's mine!"

Then there's the wonder of my eyes,
Where hills and houses, seas and skies,
In waves of light converge and pass,
And print themselves as on a glass.
Line, form and color live in me;
I am the Beauty that I see;

Ah! I could write a book of size,
About the wonder of my eyes.

What of the wonder of my heart,
That plays so faithfully its part?
I hear it running sound and sweet;
It does not seem to miss a beat;
Between the cradle and the grave,
It never falters, strong and brave.
Alas! I wish I had the art,
To tell the wonder of my heart.

Then oh! But how can I explain,
The wondrous wonder of my brain?
That marvelous machine that brings,
All consciousness of wonderings;
That lets me from myself leap out,
And watch my body walk about;
It's hopeless-all my words are vain,
To tell the wonder of my brain.

Come, let us on a seashore stand,
And wonder at a grain of sand;
And then into a meadow pass,
And wonder at a blade of grass;
Or cast our vision high and far,
And thrill with wonder at a star;
A host of stars-night's holy tent
Huge, glittering with wonderment.

If wonder is in great and small,
Then what of Him who made it all?
In eyes and brain and heart and limb,
Let's see the wondrous works of Him.
In house and hill and field and sea,
In bird and beast and flower and tree,
In everything from sun to sod,
The wonder and the awe of God.
Adapted from Robert Service

Reference:

White, F. N. 1977. Circulation. In *Animal Physiology: Principles and Adaptations* by M. S. Gordon in collaboration with G. A. Bartholomew, A. D. Grinnell and F. N. White. Pages: 218-243.

Hummingbirds hibernate at night

Are not two sparrows sold for a penny? Yet not one of them will fall to the ground apart from the will of your Father. (Matt 10:29)

Ruby-Throated Hummingbird *Archilochus colubris*
Photograph by Robert Lubeck. Source
http://animals.nationalgeographic.com

Hummingbirds have always impressed my family. My grandmother planted honeysuckles near the house to attract them. I remember watching them as a kid and was sometimes able to get close enough to hear the distinctive hum of their rapid wing beats. I was impressed how they could hover near a flower and even fly backwards after harvesting the nectar. Later as an adult, I put up hummingbird feeders and still enjoy watching them. Their small size and helicopter-like flight have always fascinated me, especially after getting my pilot license and owning an airplane.

Some readers will find this disturbing, but I remember my grandfather wanted to see if they were actually birds or just another kind of sphinx moth (we call them "hummingbird moths" in Oklahoma). He shot one with a 0.22 rifle and knew first hand that they were actually birds with tiny feathers. I was also impressed that he could hit such a small moving target. Please understand this was long before our birds were protected by law. Back then, country folks depended on hunting for food and the trusty 0.22 was always in reach. I still own the same 0.22 rifle Grandpa used. It was manufactured in 1911.

As a zoologist I have been impressed by several aspects of these marvelous little birds. Let me share some of the things that make them unique. First, they are indeed the smallest bird and are the smallest of all warm blooded or endothermic animals. The shrew is next and is not much bigger. Yes, I have even found a couple of their little nests. They line them with soft spider webs to protect their delicate pea sized eggs. As one might expect due to their small size, hummingbirds in flight have the highest rate of metabolism of all animals except a few flying insects. Their heart rate can reach 1,260 beats per minutes. Even at rest, the have an extremely high rate of metabolism and are always but a few hours from starving to death. It is uncanny that the tiny ruby-throated hummingbird is able to migrate 800 KM (500 miles) across the Gulf of Mexico twice a year. There is another more frequent problem.

They barely have the ability to store enough energy to stay alive on warm nights. In order to survive long cold nights they must reduce their rate of metabolism even farther especially when food is in short supply. They do this by lowering their body temperature. In other words they go into nightly hibernation more accurately known as torpor. In this state, their heart rate slows to 50 to 100 beats per minute and their body temperature drops to near the ambient temperature as is the case for hibernating mammals. While this solves the problem of starvation at night it creates another problem for females incubating eggs. The time required for the incubation of most birds is well known and fixed. With hummingbirds it is highly variable. The reason is simple. If it is cold and the bird must go into nightly torpor it takes longer for the eggs to hatch. Incubation typically takes from 14 to as long as 23 days depending on the weather.

Once again this presents a plethora of problems for the evolutionist. Since it is so effective for hummingbirds, why has it not evolved for other endothermic animals? It requires many special enzymes and other compounds for neurons to operate over the wide temperature ranges. Once again the behavior, physiology and biochemistry must all be present at the same time for them to accomplish these remarkable feats. It is impossible to conceive of this complex response developing in small steps caused by mutations or genetic errors. Once again evolution has failed to explain the facts of science.

References

Chambers, Lanny. "About Hummingbirds". Hummingbirds.net. http://www.hummingbirds.net/about.html#heartbeat. Retrieved 25 January 2009.

Hainsworth, Reed; Wolf, Larry (May 1993). "Hummingbird Feeding". *Wildbird Magazine*. http://www.hummingbirds.net/hainsworth.html.

Suarez, R. K.; Gass, C. L. (2002). "Hummingbirds foraging and the relation between bioenergetics and behavior". *Comparative Biochemistry and Physiology*, Part A. 133: 335–343.

Warrick, D. R.; Tobalske, B.W. & Powers, D.R. (2005). "Aerodynamics of the hovering hummingbird". *Nature* 435: 1094–1097

Nutria's mammary glands are on their back

Nutria, *Myocastor coypus*
Nationalgeographic.com

Nutria are common where alligators live and I enjoy watching them. They are sometimes called swamp Beaver or Nutria rat. This semi-aquatic rodent was introduced from South America in the 1930's for use as a fur bearing animal. There were many nutria farms, but after the fur market collapsed in the 1940's, many of the animals were released. They have become a common nuisance in waterways throughout the southeast. They are large stocky built animals living around water and are often confused with the larger beaver. Unlike beaver with their distinctive large flat tail, nutria has a small round hairless rat-like tail. Adults weight from 25-30 pounds and are dark brown in color. The feed on plants such as duckweed in the water and on land and have been seen grazing on grass with cattle. In marshy regions they construct a platform out of nearby plants on which to eat. Those living in other areas have an underground borrow.

They spend a great deal of time in the water and have a unique feature not widely known. Their nutria's mammary glands are actually located on their backs so their young can nurse while the mother is swimming and searching for food. Once again the actual facts of science produce a paradox for evolutionists. Obviously there are many semi-aquatic mammals in the world, yet this unusual anatomy is found only in nutria. It is not a good time to be an evolutionist.

References
Adams, W.H. 1956. The nutria in coastal Louisiana. Louisiana Academy of Sciences 19:28-41.

Atwood, E.L. 1950. Life history studies of nutria, or coypu, in coastal Louisiana. Journal of Wildlife Management 14:249-265.

Ehrlich, S. 1966. Ecological aspects of reproduction in nutria *Myocastor coypus* mol. Mammalia 30:142-152.

Wade, D. A. and C. W. Ramsey. 1986. Identifying and Managing Aquatic Rodents in Texas: Beaver, Nutria and Muskrats. Texas Agriculture Extension Service. 46 pp.

Ducks have cold feet

All flesh is not the same: Men have one kind of flesh, animals have another, birds another and fish another. (1 Cor 15:39)

Stately Mallard duck
animal.discovery.com AP photo

One of the reasons I like zoology is the endless array of special features that enables them to survive in diverse and often harsh environments. Ducks provide an excellent example. Certainly their annual migration over thousands of miles is impressive, but I am more in awe at their feet. Certainly the webbing between the toes provides excellent propulsion for swimming, but there is more. Ducks return to their summer nesting areas when it is still cold at night and the water is near freezing. They spend most of their time swimming in icy cold water. A duck's feet are relatively large and comprise a significant percentage of their total surface area, yet they lose very little body heat. How is this possible? If our hands or feet are in ice water for even a short time we begin to feel cold all over because of the heat loss. Certainly a duck's body is covered with warm and water repellant feathers and down, but not their feet and their feet are in direct contact with ice cold water. Obviously a rich supply of blood is required to nourish and provide oxygen for the skin, nerves and muscles of their feet. Why do the feet not lose a significant amount of heat to the cold water? The answer is easy to comprehend, but presents yet another difficult conundrum for evolutionists to explain by random mutational genetic errors over time.

The feet of ducks have a simple, but highly efficient heat exchanger. The warm descending arterial blood from the heart is in close contact with the cold ascending blood from the vein that returns blood from the heart. The blood vessels break up into small vessels that intertwine with each other. The blood is flowing in opposite directions and the warm arterial blood is cooled by the cold blood from the foot. The cold blood from the feet is warmed from the arterial blood and little heat is lost or wasted. The technical name for this efficient heat exchanger is tibiotarsal or simply the arterio-venous association.

There are other examples of counter current exchange systems in living things. The evidence is in and the conclusion is obvious. Ducks were designed to survive and thrive in ice cold water while maintaining their high and stable body temperature. There is not other rational explanation. The next time you see ducks or geese swimming in ice cold water give honor where honor is due…to the One who designed their efficient yet simple heat exchanger in their feet.

References

Kilgore, D.L. & Schmidt-Nielsen, K., Heat loss from ducks' feet immersed in cold water, The Condor, 77:475-517, 1975.

Midtgard, U., The rete tibiotarsale and arterio-venous association in the hind limb of birds: a comparative morphological study on counter-current heat exchange systems, Acta Zoologica, Vol. 62, No. 2, 67-87, 1981.

Wolves turn the other cheek

But ask the animals, and they will teach you, or the birds of the air, and they will tell you; or speak to the earth, and it will teach you, or let the fish of the sea inform you. (Job 12:7-8)

Gray Wolf
animals.nationalgeographic.com

The Bible instructs us to study the world around us and to ask questions of living things. The above scripture passage was one of my motivations for becoming a zoologist. I have always seen God in His Creation as has been the case for many other scientists in the past. Here are a few well known scientists who saw God's hand in nature. Many more examples could be given.

Johannes Kepler was an outstanding German mathematician and astronomer. It was he who discovered that the earth and other planets travel around the sun. He described his own discoveries as "thinking God's thoughts after Him."

Albert Einstein said, "Everyone who is seriously interested in the pursuit of science becomes convinced that a spirit is manifest in the laws of the universe--a spirit vastly superior to man, and one in the face of which our modest powers must feel humble."

Blaise Pascal was a French mathematician, scientist and philosopher. He invented the first working barometer and made such significant innovations in the field of probability science and mathematics that today a computer language is named after him. In his book, *"Pensees,"* he wrote, "Faith tells us what the senses cannot, but it is not contrary to their findings. It simply transcends, without contradicting them."

Sir **Isaac Newton**, mathematician and physicist, is considered by many to be one of the foremost scientific intellects of all time. In his work titled, *"Principia"* he said that "This most beautiful system of the sun, planets, and comets, could only proceed from the counsel and dominion of an intelligent and powerful Being...Atheism is so senseless. When I look at the solar system, I see the earth at the right distance from the sun to receive the proper amounts of heat and light. This did not happen by chance."

Astronaut Jim Irwin was so moved by the presence of God he felt as he walked on the moon, that he resigned from the astronaut program to become an evangelist.

We are to study these things not only to better understand the world that God made, but also as examples life issues we need to learn. Two familiar teachings of Jesus illustrate this important principle. *Are not two sparrows sold for a penny? Yet not one of them will fall to the ground apart from the will of your Father.* (Matt 10:29) If God determines the life and death of an insignificant sparrow, how much more is He involved in our lives? Unlike the animals, we alone were created in His image. Consider the plants. *And why do you worry about clothes? See how the lilies of the field grow. They do not labor or spin. Yet I tell you that not even Solomon in all his splendor was dressed like one of these.* (Matt 6:28-29)

Jesus taught another important life principle in the following passage, yet it is often misunderstood today. But I tell you, do not resist an evil person. If someone strikes you on the right cheek, turn to him the other also.

And if someone wants to sue you and take your tunic, let him have your cloak as well. If someone forces you to go one mile, go with him two miles. (Matt 5:39-41) There is a little known aspect of wolf behavior that helps us understand the deeper meaning of this important teaching.

The year was 1980 and I was keynote speaker at a major international telemetry conference at Oxford University in England. I had three of my research students with me. The BBC TV documentary, *A smile for the Crocodile* had just been released in the United States and in Europe. The film featured my alligator research and people at the meeting and even on the streets of London approached me and said, "You are Dr. Smith and you study alligators." At the conference I met Dr. David Mech. Even then he was a wolf expert. He respected my work with alligators and we chatted for some time. I spoke to him recently by phone and he still remembered me from the conference all those years ago.

Today Dr. Mech remains an internationally recognized wolf expert and senior research scientist for the U.S. Department of the Interior and has held that post since 1970. He also teaches as an adjunct professor at the University of Minnesota. He has studied wolves in their natural habitat since 1958 in Minnesota, Yellowstone National Park, Alaska, Canada, and Italy as well as in other places wolves survive today. He has published ten books and numerous scientific articles about wolves and other wildlife, the best known of these are *The Wolf: The Ecology and Behavior of an Endangered Species* (1970, University of Minnesota Press) and *Wolves: Behavior, Ecology, and Conservation* which he co-edited with Luigi Boitani (2003, University of Chicago Press). At that conference long ago he shared something that illuminated not only wolf behavior, but also provides insight on that teaching of Jesus that had troubled me for many years. It is simple, yet deeply profound. Let me explain.

A wolf "turns the other cheek" to its enemy

Wolf packs usually stay within well marked home ranges. Under normal conditions the home range provides ample living space and food for the pack. Occasionally, during difficult years, there are battles between packs to expand their territory, but these are rare. More common are fights within the pack between adult males for food or a female. The stakes are high; the battle intense. Like most predators, they have overkill potential and can kill animals many times their size. They could easily kill another wolf, yet a curious behavior allows the loser to escape unharmed. After an intense struggle the beaten wolf does a strange and seemingly unwise thing. He rolls over on his back and reveals his most vulnerable area...his throat. In a very real sense he is "turning the other cheek." This submissive posture by the one who is attacked totally inhibits farther aggressive behavior from the attacking wolf. Peace returns to the wolf community.

So it is with human behavior. If someone abuses you verbally or in other ways our human nature is to fight back. Our sin nature demands, "An eye for an eye." An argument ensues and the situation can rapidly escalate. This is our human nature and we all understand it. The next time this happens, consider applying the profound teaching of Jesus. Instead of striking back, turn the other cheek or "bury the hatchet" by apologizing or complimenting the enemy. A surprising thing often happens. As with the fighting wolves, farther escalation and aggressive behavior is suddenly and irreversibly inhibited.

Let me give an example that has grown to have deep meaning to me over the years. My varied work history includes a stint as an electronic technician at a consumer products (radios and TV's) application laboratory at Texas Instruments near Dallas, Texas. As often happens in industry, we had an excellent manager. Everyone liked him. Sadly, he was promoted to a higher management level. We all knew the previous boss left some large shoes to fill. Soon his replacement arrived. We had heard the scuttlebutt and it was not good. We were apprehensive.

The new boss arrived and during the first week, he made it a point to have a little "one on one time" with each member of his new team. What he did was very much like turning the other cheek. After chatting with each of us alone for a short time about what we did and our future aspirations with the company, he did something that at the time I thought strange. In hindsight, I see it as genius. He found something about each one of us that was an important part of who we were and complimented us on it. At the time I was the only technician sporting a beard. He complemented me on my beard. No one had ever done that before and certainly not a supervisor. Needless to say I was instantly on his side. I liked him. He had disarmed me and I could find no fault in him. He did the same with each of the other team members and the transition to a new leader was smooth and we liked him even more than the previous team leader.

It seems there are not one but two lessons here. If someone insults us or hurts us either intentionally or by a misunderstanding, we MUST fight our human nature, our fleshly desire to retaliate and escalate the situation. Instead, we should think of something kind to say; offer a compliment instead of an insult. In doing so we are turning the other cheek and in the process can turn a potential enemy into a friend. It works for wolves and seems to be human behavioral trait as well.

In a similar way, when we meet someone for the first time we need to spend a bit of time getting to know them. Yet there is more we need to do. In getting to know them we must also find something that is important to them and find a way to compliment them on this trait. In today's jargon I see this as "pro-active cheek turning." Jesus taught it and we should follow the principle with confidence. I learned this important life principle from one of North America's most majestic creatures, the wolf, ***Canis Lupus.***

Counter current flow in fish gills

Herring
NOAA (National Oceanic and Atmospheric Administration)

Perhaps the best example of the counter current blood flow system mentioned above in the "Ducks have cold feet" chapter is found in the gills of fish. All animals require oxygen to survive. It is much easier to get oxygen from air than from water. The air we breathe contains 21% oxygen while the amount of oxygen in water is often less than 1%. Water is much denser than air and harder to move. Fish were designed with a highly efficient counter current blood flow arrangement similar to that seen in the in the feet of ducks to conserve body heat.

The actual gas exchange takes place in tiny gill filaments or lamellae in a fish's gills. The lamellae are very thin walled and are only 1-5 microns thick. The number of lamellae increases with the body size of the fish. It has been estimated that for a body size of 1 KG (2.2 pounds) the fish gills contain 5 million lamellae for gas exchange. Here the water and blood flow in the opposite direction. It is also noteworthy that water flows not stop in the same direction through the fish's gills. In other air breathing animals they inhale, stop and exhale. This also makes fish gills a much more efficient method of getting oxygen and getting rid of carbon dioxide.

Water from the stream, pond or lake first encounters blood that has already taken on oxygen from the water and yields the maximum transfer of oxygen. As the water continues to flow alongside the gill filament it is adjacent to blood containing less and less oxygen providing the highest possible transfer. This method removes up to 90% of the available oxygen from the water. Experiments have been performed by surgically reversing the direction of the blood flow. It was discovered that when the blood and water flow in the same direction only about 50% of the oxygen could be transferred from the water to the blood.

Once again we see evidence of design in the way fish exchange gas. As we have seen before such a system must have all the parts arranged in the correct way before any advantage is seen. Such a complex system of gas exchange could not develop by small mutational errors. Again, evolution has failed and diehard evolutionists do not even have a rational theory to explain such a beautiful and efficient design as seen in the countercurrent system seen if a fish's gills.

References

Levesque, M., Fralick, L., and McDowell, J. (1999). *Respiration in water: An overview of gills.* See also: www.unb.ca/courses/biol4775/SPAGES/SPAGE13.HTM

Smith, L.S. (1982). *Introduction to Fish Physiology.* TFH Publications Ltd, New York.

Wedemeyer, G.A., Meyer, F.P., and Smith, L. (1976). *Diseases of fish: Environmental stress and fish diseases.* TFH. Publications, Inc. Ltd, New York.

Fish, a quick course on Ichthyology. (1999). **Fishes - How fish breathe.** http://www.marinebiology.org/fish.htm

Hoar, W.S. and Randall, D.J. (1984). *Fish physiology Vol.X. Part A.* Academic press, New York.

Hoar, W.S. and Randall, D.J. (1970). *Fish physiology Vol.I.V.* Academic press, New York.

Which prey do predators really eat?

A lion, mighty among beasts, who retreats before nothing. (Prov 30:30)

Mountain Lion
Photo by K. Fink, NPS

Evolutionists have long held that predators preferentially take the young, weak and diseased prey. This concept is central to natural selection and is one of the tenets on which evolution rests. The premise is flawed. The entire superstructure built on natural selection providing a mechanism for evolution collapses into disarray if predators do not actually take the weakest individuals. Upon close examination the thesis is neither logical nor supported by the scientific evidence. This is an important issue because it means evolution based on natural selection lacks an important mechanism. Natural selection combined with genetic mutations is the foundation on which much of evolution rests for it is thought to provide the mechanisms by which a species can change, adapt and improve over time. For over 150 years evolutionists have maintained predators capture the weak, young, and diseased prey thereby "improving" the gene pool. Educational programs on public television such as the popular Nature series have replaced reading science material for many people. This beautifully photographed and professionally written series have aired over 400 programs since it began in 1982. Many of the episodes repeat the mantra that predators can only capture the weak. We are repeatedly told predators perform the crucial task of allowing only the fit prey survive and reproduce. By removing the weakest individuals, the predators are thought to power the evolutionary process. This selection for the "fittest" is said to be the driving force for evolution. Without predators harvesting those less fit, evolution is a theory without a mechanism, an idea without scientific merit. But is it real? Do the scientific data support this scenario?

Even as a naïve high school student, I scoffed at this idea upon first hearing it in biology class. Young animals are only available during a small fraction of the year and most wild animals are healthy. If predators had to rely on eating young or sick prey they would quickly starve to death. There is another fundamental problem with this theory. If predators ate diseased animals they would likely become ill. This is common sense and has been known since at least the time of Moses: *Anyone, whether native-born or alien, who eats anything found dead or torn by wild animals must wash his clothes and bathe with water, and he will be ceremonially unclean till evening; then he will be clean* (Lev 17:15). People of all cultures learned to avoid eating sick animals or those that have died of disease. As I matured, became educated and studied wild animals in the field under natural conditions, the concept of predators only being capable of catching the weak prey become more and more improbable. Let's consider some of the reasons this theory lacks credibility and see if it is supported by scientific evidence.

Most predators have overkill potential. As is often seen on public television Nature programs a cheetah or other cats are capable of catching, killing and eating prey larger than they are. If you have seen a cat catch a mouse or a dog chase a cat or rabbit, you know the chase-kill instinct is powerful driving force for many predators. In a 20 year study in New Zealand, it was demonstrated that well fed farm cats would travel 3 km to kill wild rabbits (Gibb, et. al, 1978).

Death feigning

One powerful argument that predators are not looking for an easy meal is death feigning widely seen in a large number of animals. If predators were looking for an easy meal, for the prey to drop to the ground and feign death rather than running or hiding seems suicidal. Yet a number of animals take this approach when attacked by a predator and it therefore must provide some level of protection. When frightened or injured the Eastern Hog-nosed snake, *Heterodon platyrhinos* rolls over on its back and feigns death. In an almost comic fashion, if you roll it over in the normal position, it immediately rolls back over on its back. The message seems to be, "To properly "play" dead you MUST be lying on your back" (Burghardt and Greene, 1988).

Ornate box turtle, Terrapene ornata
(Photo by Sean Williams)

There is another example of death feigning that I remember from my childhood and studied later as a

194

physiologist. Ornate box turtles, *Terrapene ornata* are common in the pastures and gardens of western Oklahoma where I grew up and still live. Certainly, their strong shell provides protection from most predators, but there is more. Their behavior associated with being threatened by a predator also has survival value. In addition to their protective shell, when disturbed they pull their head and feet inside and remain motionless. They feign death as do a number of animals (Smith and DeCarvalho, 1985). Upon finding a box turtle moving slowly along a dog will approach it and attempt to bite it or scratch at it with its feet. The turtle promptly retreats to the safety of its shell, closes the hinged front portion of the plastron and remains motionless. It feigns death and is inaccessible. Soon the dog or other predator loses interest in the non-responsive turtle and moves on in search of more challenging prey. Death feigning seems to be a common response among animals and as we will see below sometimes occurs in humans.

When frightened by an approaching predator many animals seek refuge in a safe hiding place. This passive fear response is equally widespread, but less well known than the classic fight or flight response. Such hiding animals remain motionless and reduce their metabolism resulting in a marked reduction in both respiration and heart rate. Body temperature may also drop. Unlike the sympathetically dominant fight or flight response this passive response is parasympathetically dominant and reduces the likelihood of being detected and killed by a predator.   The response has been described for every major group of vertebrates including man (Smith, 2006). One can only conclude such a widespread and profound physiological response must have high survival value (Honma, *et al*, 2006).

Death feigning in the American Opossum
Photo by author

There are variations in the details of how various animals respond to fear by hiding and remaining motionless. Perhaps the best death feigning actor is the American Opossum, *Didelphis virginiana*. Their heart rate drops over 50% when feigning death and they are totally unresponsive to touch. Even the cornea of the eye can be touched with the normal blinking reflex. In spite of this appearance they are fully conscious. When the predator retreats their heart rate gradually returns to normal. If the predator returns they will again reduce their heart rate even if they are not touched by the predator clearly demonstrating they are conscious and aware of their surroundings (Gabrielsen and Smith, 1985).

The opossum's death feigning performance has earned them a popular phrase in the American English language. People are said to be "playing 'possum" when unresponsive to events around them. There is a similar and even broader term we sometimes hear, that of being "paralyzed by fear." This is another manifestation of the death feigning response and also provides a high level of protection from predators.

Scripture provides an excellent example of this response from a most unexpected source. Many agree the Roman solders were the best trained and most disciplined warriors at the time. Yet, upon witnessing the bodily resurrection of Jesus Christ, even these veteran fighters were paralyzed by fear and feigned death. *The guards were so afraid of him that they shook and became like dead men* (Matt 28:4, NIV). So it is with many animals. Death feigning is a widespread response to attack or even the approach of a predator. How can such a widespread response have survival value?

Chase-kill sequence

As any dog owner knows, dogs enjoy chasing things from chew toys to the neighbor's cat to automobiles. So it is with most predators. They seem to enjoy the chase-kill sequence. Let me give some examples to illustrate this important but poorly recognized aspect of predator behavior.

It is common knowledge among herpetologists that it is difficult to get captive snakes to eat food they have not killed. For example pythons will sometime go months before they will accept dead prey. I had a pet boa constrictor for 23 years and often fed it fresh road killed rabbits. However in order to get him to take the road kill, I had to warm the dead rabbit in the microwave and then move it inside its cage before it would strike. Boas have labial heat sensors and prefer warm prey. Such instinctive behavior helps many animals avoid eating dead prey that could make them sick.

There are exceptions. Vultures are known to eat animals that die of natural causes as well as road kill. Their stomach acid is exceptionally corrosive enabling them to digest putrid carcasses infected with botulism and other bacteria lethal to other scavengers. Hawks, opossums and a few other animals are also known to eat carrion without ill consequences.

While completing my doctoral research with alligators at the Welder Wildlife Refuge in south Texas I met graduate student, Roy McBride. He was older than most students and had an uncanny skill. He could track cougars or mountain lions better than anyone. Prior to becoming a graduate student he had worked as a bounty hunter tracking and killing nuisance mountain lions that killed livestock throughout the southwest and Mexico. He could recognize which individual cat had made the kill by careful examination of the carcass. Each lion had individual preferences. Some preferred internal organs like the liver or heart which they would eat it first. Others preferred muscle. He could also tell a lot about what the mountain lion was doing by following its tracks. For example, if a lion was simply moving from one area to another it would follow low lying areas and remain out of sight. If instead, it was hungry and looking for prey, it would move from one high look out area to another scanning the surroundings looking for something to eat. There were four species of cats on the refuge: mountain lions *Felis concolor,* bobcat *Lynx rufus*, Jaguarondi *Puma yagouaroundi* and an occasional ocelot, *Leopardus pardalis*. Roy McBride could distinguish between their footprints and track all of them. His focus at the refuge and for his Master's thesis research involved the largest of these four, the mountain lion as well as its behavior and feeding habits.

He related an observation to me that bears directly on this discussion. He was tracking a large mountain lion in south Texas and it was hungry and looking for something to eat. He knew this because it was moving from one lookout place to another searching for prey. During its hunt, the hungry predator came across a live deer with its antlers tangled in a fence. The tracks revealed the lion approached the deer first from one side then the other, but moved on searching for other prey. If it were looking for an easy meal as evolutionists would have us believe, it would have killed and eaten the entangled deer, but it did not. Details of his study and other mountain lion observations are reported in detail in his Master's Thesis (McBride, 1977).

He also shared another telling example that occurred several times in Mexico. He worked with ranchers, again protecting the herd from predatory mountain lions. In this area of Mexico cattle are taken to market only once a year. Some of the younger calves were weaned very young and had difficulty keeping up with the herd. They often straggled behind, making easy targets for the mountain lions. Without fail the lions ignored the young weak calves, but instead attacked and killed the large healthy 500-600 pound steers. Once again this demonstrated the fallacy in thinking these predators select the weak and flies in the face of evolution dogma. I recently confirmed these accounts with Mr. McBride.

He told me of another incident he observed. Working in Florida with sheep farmers he has developed a collar that releases a poison to kill the mountain lion or other predator that attacks lambs. The ranchers did not want to sacrifice their strongest lambs and had him place the collars on the weakest and smallest lambs. Without exception, the lions left these animals alone and sought out and killed larger healthy lambs. In order to control these predators the ranchers allowed him to install the protective collars on their largest and healthiest lambs.

McBride has continued his research in Texas with similar results (McBride, 2003). We have been misled. Predators are not looking for an easy meal as evolutionists would have us believe. They prefer and seem to need the chase-kill sequence.

Certainly other factors are involved in determining which individual prey animal is taken by a predator. Some of the smaller predators may indeed, select smaller individuals. Other predators may be opportunistic and take an individual that was simply in the wrong place at the wrong time. Still, the above observations are important and scientists need to know more of the details in what determines which individual prey is sought out and killed by various kinds of predators. The answer does not seem to be as simple or as clear cut as evolutionists have been saying. Additional research is sorely needed in this important area.

Conclusion

Observation clearly show predators do not consistently select the weak, sick or young as evolutionists have long accepted and taught. Many predators have overkill potential and can easily catch and kill larger healthy prey. Predators also seem to seek the chase-kill sequence and will actually ignore live prey that will not flee when approached. Feigning death by the opossum and other animals provide strong evidence that something is amiss with the current view. The entire predator/prey relation needs to be studied in depth and re-evaluated. It appears the evolutionists have been misled and one of their important foundation cornerstones is cracked and will soon disintegrate.

References

Burghardt, G. M. and H. W. Green. 1988. Predator simulation and duration of death feigning in neonate hognose snakes. **Animal Behavior 36**:842-44.

Gabrielsen, G. W. and E. N. Smith. 1985. Physiological response associated with feigned death in the American Opossum. **Acta Physiol. Scan. 123**:393-398.

Honma, A, O. Shintaro and T. Nishida. 2006. Adaptive significance of death feigning posture as a specialized inducible defense against gape-limited predators. Proc.Biol Sci. July 7; 273 (1594): 1631-1635.

Gibb, J. A., P. C. Ward and G. D. Ward. 1978. Natural control of a population of Rabbits (Oryctolagus cuniculus (L) for ten years in the Kourarau enclosure, 88pp. D.S.I.R Bulletin 223, Wellington, NZ.

McBride, R, T. 1977. Status and Ecology of mountain lions, *Felis concolor* of the Texas-Mexican border. Master's Theses, Sol Ross State University, Alpine, TX.

McBride, 1980. Report on Mountain Lion Survey, Guadalupe Mountains National Park. National Park Service Special Report, pp3.

McBride, R. 2003. The effects of Predator Control on Mountain Lions in Texas. Page 72 in L. A. Harveson, P. M. Harveson and R. W. Adams eds *Proceedings of Sixth Mountain Lion Workshop,* Austin, Texas.

Smith, E. N. and M. C. DeCarvalho, Jr. 1985. Heart rate response to fear and diving in the ornate box turtle, *Terepene ornata. Physiol. Zool.* 58:236-241.

Smith, E. N. 2006. Passive Fear: Alternative to Fight or Flight. Published by **iUniverse**, New York

Smith, E. N. 2010. Which prey do predators eat? *Australian Journal of Creation* Vol. 24(2).

Conclusions and application

Have nothing to do with godless myths and old wives' tales; rather, train yourself to be godly.   For physical training is of some value, but godliness has value for all things, holding promise for both the present life and the life to come. This is a trustworthy saying that deserves full acceptance.
(1 Tim 4:7-9).

Certainly the material in this book will do little to change the minds of die hard evolutionists, but for others the information will be useful. It is hoped the material presented here will confirm those already doubting the validity of evolution for it is clearly NOT the panacea it was once hoped to be. Careful examination of the actual scientific evidence clearly shows evolution has failed not only to account for the origin of life, but also as an explanation for the complexity and diversity seen in all living things. Even the fossil record belies the gradual transformation of ANY major kind of animal into another. Science students in particular MUST have the freedom to examine and openly discuss the evidence and follow were it leads...even if it leads to Intelligent Design and the Creator. Science must again be objective...without it, science as we know it will die.

There is also a need for Christians to understand religious persecution remains rampant today in the hallowed halls of academia. Anyone critical of evolution or Darwin is at great risk of being denied tenure or fired. This too must stop. We need both freedom of speech even when it is contrary to the accepted view. Science has been wrong in the past and will be wrong again in the future. I strongly feel it is wrong today regarding evolution. Only by careful examination of the facts and open debate can progress be made. It is hoped scientists and especially the new generation of science students will be free to examine the evidence and discuss the pros and cons openly in biology classrooms. Fraud and dishonesty in textbooks and lectures MUST stop. No one should fear the truth...no matter where it leads.

Meet the Author

Doc Gator with a soul mate

E. Norbert Smith has an earned doctorate in zoology from Texas Tech University, a Master's degree in biology from Baylor University and a bachelor's degree in biology from Southwestern Oklahoma State University. He has caught, studied and released over 200 alligators up to 750 pounds and still has all his body parts. That is why his students called him, Doc Gator.

He designed and used sophisticated multichannel radio telemetry to study behavioral and physiological thermoregulation of alligators and the cardiovascular response of wild animals to fear. He put heart rate transmitters on more species of wild animals than anyone in the world. His work was featured in the BBC TV documentary, *A smile for the Crocodile* and he was invited as keynote speaker to a major international radio telemetry conference at Oxford University. He was awarded the "Outstanding Teacher" award two years and was highly successful getting his students into medical and graduate schools. He had over 50 research papers and abstracts published in leading scientific journals during his five years at Northeastern Oklahoma State University. Yet he was denied tenure for his rejection of evolution thus ending his professional career. He worked as an oil field roughneck and became a truck driver until his retirement six years ago. He enjoys retirement and still lives on the family farm in western Oklahoma. He enjoys gardening, nature photography and writing. He is currently working on several Bible related articles and books.

He has published nearly 100 technical publications as well as numerous popular electronics articles. He has published children's articles in **Ranger Rick** and **Highlights** magazines. He has also published six children's books about Al-the-Gator and his friends in a farm pond in South Texas. Several more have been written and are awaiting the artwork. He is active in Creation science and has published articles in the **Creation Research Society Quarterly** since his undergraduate days. He served on their board of directors for the Creation Research Society and taught a graduate course for the Institute of Creation Research as well as an online course in Creation for Liberty University. He recently published a major article in the

Australian *Journal of Creation*. He is currently working on papers dealing with the mechanism of increased pain in childbirth as well as several other Bible related articles and books. He recently published two major books that might be of interest to readers of this one. They are available online or at your local book store.

**A major scientific apologics book
in defense of Biblical Creation**

Synopsis: Central to the book are over 1,500 Bible verses related to God as Creator and Sustainer. Evidence for and against evolution are given and details of modern religious persecution with over 3,000 professors being denied tenure or fired for doubting evolution. A must read in these increasingly anti-Christian times.

Scientific apologetics book revealing the lack of objectivity in modern science

Synopsis: The term "sacred cow" is fitting as part of the title for this book and comes from the veneration of cows by the Hindus of India. As early as 1910 the term was used as a metaphor to describe a person, organization or institution that is unreasonably immune from criticism. It is fitting when describing certain aspects of macroevolution, but it also applies today for other areas of science. In the past science was defined largely by its method. Carefully controlled experiments, provisional conclusions, and considered debate once defined the field. But those days have passed. Today, science is often defined by public policy statements, consensus, and a set of metaphysical assumptions that cannot be directly tested. We are taught that science is above all things objective. Unfortunately this is no longer true today as we attempt to show in this groundbreaking book.

AL-THE-GATOR AND SNEAKY SNAKE

by E. Norbert Smith, Ph.D.
Illustrated by Linda Moore

This is my fifth children's book. Each one has a moral lesson from the story and a list of additional reading for kids who want to dig deeper. There are a few Bible verses about Creation and some interesting facts about the animals. It is unusual for the author of children's books to also be a scientist. I have studied most of the characters in the books I write.

Dr. Smith enjoys retirement and has started writing his twenty-eighth book. Visit his websites: www.GodofCreation.com and www.NorbertSmith.com. No doubt with any project of this magnitude there will be errors and omissions. Please let me hear from you with comments…good or bad. Contact him via email:

DocGater@aol.com or snail mail to: E. Norbert Smith, Ph.D, 24340 East 1080 Road, Weatherford, OK 73096, USA

www.ingramcontent.com/pod-product-compliance
Lightning Source LLC
Chambersburg PA
CBHW071420170526
45165CB00001B/337